Metaphysics of Infinity

The Problem of Motion and the Infinite Brain

Ion Soteropoulos

UNIVERSITY PRESS OF AMERICA,® INC.
Lanham • Boulder • New York • Toronto • Plymouth, UK

University Press of America,® Inc.
4501 Forbes Boulevard
Suite 200
Lanham, Maryland 20706
UPA Acquisitions Department (301) 459-3366

10 Thornbury Road
Plymouth PL6 7PP
United Kingdom

Printed in the United States of America
British Library Cataloging in Publication Information Available

Library of Congress Control Number: 2013938015
ISBN: 978-0-7618-6146-1 (clothbound : alk. paper)
eISBN: 978-0-7618-6147-8

Cover image by Setareh Korkchi and Ion Soteropoulos.

∞™ The paper used in this publication meets the minimum requirements of American National Standard for Information Sciences—Permanence of Paper for Printed Library Materials, ANSI Z39.48-1992

I dedicate this book to the lovers of divine Logos

To know the physical thing completely
we need at most two things
united by an intimate bond.
The couple forming a unitary whole
is therefore
the first origin and principle of things
whose infinite power of two
in doubling and dividing
determines the magnitude of the world.

—Ion Soteropoulos

Contents

List of Figures

Preface

Since the time of the Greek philosopher Zeno (fifth century BCE), who formulated his paradoxes of motion, our faculty of analytic understanding has failed to comprehend motion through the ages. We have stated the Newtonian laws of motion and discovered the theories of relativity and quantum mechanics; we have built telescopes and microscopes that probe beyond the scales of 10^{25} meters and 10^{-15} meters; we have constructed spacecraft that travel beyond the limits of our solar system, but the nature of motion from *a* to *b* by passing through an infinite series of sub-distances in one instant remains unintelligible.

In this work, we will show how the complex idea of infinity solves the paradoxical or contradictory nature of motion and determines its founding principle. First, we will investigate Aristotle's analytic solutions as proposed in his *Physics* VI. Next, we will examine the modern analytic theory of infinite convergent series and show why analytic principles of thought, which are abstractions of our individual perception of things, can in no way help us to comprehend continuous motion from *a* to *b*. Because motion resides outside the analytic principles of our understanding, we then return to the dawn of science and philosophy in order to find fresh insights and inspiration. Based on the theory of motion as postulated by Anaxagoras (Ionian Greek philosopher, sixth to fifth century BCE), we propose a synthetic solution to the problem of motion that enables us to move from *a* to *b*, from the unlimited convergent series a_n to its limit *b*. In turn, we take this synthetic solution as the founding principle of continuous motion.

It is not sufficient to give a logical solution to the problem of motion by determining its founding principle, which emancipates motion from absurdity; we must also give a geometric solution by determining the geometric form of the container of motion, which is the physical universe.

Once the founding principle of continuous motion has been identified and the geometric form of its container is determined, we will investigate the sweeping consequences they have on the Aristotelian theory, particularly on his ontology of the physical body, according to which all bodies are simple and finite and therefore no finite body can travel and count an infinite number of distances in finite time and with infinite energy and force. We will discover that if we do succeed in traversing the infinite series and reach the limit *b*, under all conditions regardless of whether the infinite series is potential (Aristotle's thesis) or actual, it is because our body—our brain—is in its deepest and highest reality a complex physical whole or universe, that is both finite and infinite without contradiction or paradox.

If our finite brain is infinite, how is it possible that the infinite does not destroy the finite? How is it possible that our finite brain has within itself an infinite power source without being immediately disintegrated by its infinite temperature? If the finite and the infinite are coexisting opposites and not contradictories, when and how can our finite brain benefit from its inner infinite power source and directly experience here and now, its infinite energy, motion, and life?

What is the impact of these disruptive ideas on our finite time-conditioned society, grounded in the Aristotelian finite analytic paradigm and organized according to Euclidean linear time? Will we continue to need our lifelike robots and their computational models? Will we progress indefinitely from *a* to *b* without ever reaching the limit *b* or will we stop at *b* not because of energy exhaustion or radical transformation into another species, but because we have found a definitive solution to the limitations of our finite time-conditioned existence in a superior order—the order of the physical universe?

If this second optimistic option is retained, then what is the geometry of the physical universe in which we experience on its limiting boundary *b* our infinite life directly, free of the contradictions, injustices and pathologies of linear time? Is the form of the physical universe closed or open, finite or infinite, or perhaps both and neither?

We will end our metaphysical investigation with a critique of the finite analytic paradigm of empirical science originating with Aristotle, whom we regard as the father of finitism and analytic logic. According to the analytic paradigm that grounds our finite individual perception of things, the observable universe is composed of simple individuals having at one time a unique determination. In this sense, inert matter is simple: that is to say, it has at one time a unique determination, for example the determination of rest; it occupies therefore at one instant a unique position—the absolute resting position *a*. This conclusion, however, contradicts the famous definition of nature (φύσις) given by Aristotle in his *Physics* II, according to which "nature is the principle of rest and motion residing immanently in the thing by essence and not by accident."

If nature is double (δύο) and divided (διχῶς), that is, both *rest*, which we designate by *a*, and *motion*, which we designate by not-*a* (*a*′) called *b*, then a physical thing, that is, a thing in conformity with nature (κατά φύσιν) is not the simple individual *a*—the indivisible point having neither parts nor motion—but rather the complex whole—the undivided line *ab* divided into the coexisting contrary parts *a* and *b* and receptive of continuous motion. It follows that insofar as matter is a physical thing, inert matter is not a simple individual, as the finite analytic paradigm asserted at the beginning, but instead a complex whole or universe having at one time contrary determinations. Thus, in Empedocles' poetic terms, "Nature is but the name that men give to mixture," whereas in logical terms, nature is the unity or coincidence of opposites (*coincidentia oppositorum*), of *a* and *b*—of rest and motion, form and matter, particle and wave, point and line, finite and infinite without contradiction or paradox.

The famous definition of nature given by Aristotle, or, more precisely, by his infinite mind (νοῦς)—in modern epistemological terms we would say by his faculty of infinite synthetic reason—contradicts the surface reality of our immediately observable world, in which things appear to be simple individuals ruled by analytic principles of organization. For example, the analytic principle of contradiction stipulating that nothing is both *a* and *b* = *a*′ violates nature's synthetic principle of unity of opposites. Is it possible to resolve this tension between our finite individual senses organizing our familiar observable world in regard to their analytic principles and the distant physical world out there, which is the real thing? Is it possible to have an infinite universal perception of things that abolishes this tension?

These are the questions we will pose throughout this work and which we attempt to answer by assuming that, in conformity with synthetic nature, every unanswerable question has simultaneously its answer.

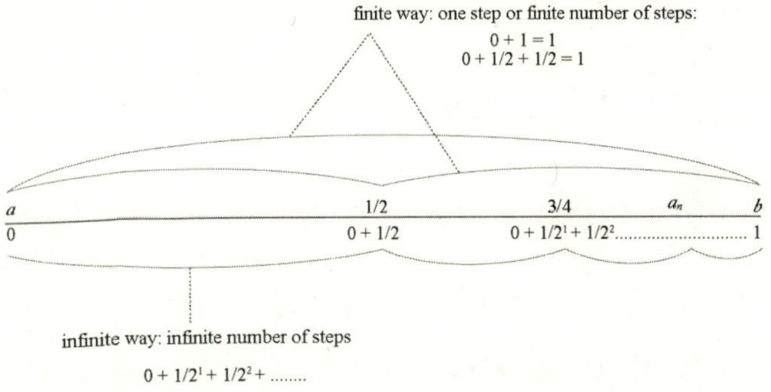

Figure 1.1. There are two ways to move along the Euclidean unit distance *ab:* i) the finite way, which consists of one step—for example, 0 + 1—or a finite number of steps—say, 0 + ½ + ½—and ii) the infinite way, which consists of an infinite number of steps—for example, 0 + $½^1$ + $½^2$ + ... Because we are finite bodies with finite energy and time, we perform actually one step or a finite number of steps and only potentially an infinite number of steps. Therefore, we access infinity only potentially and indirectly.

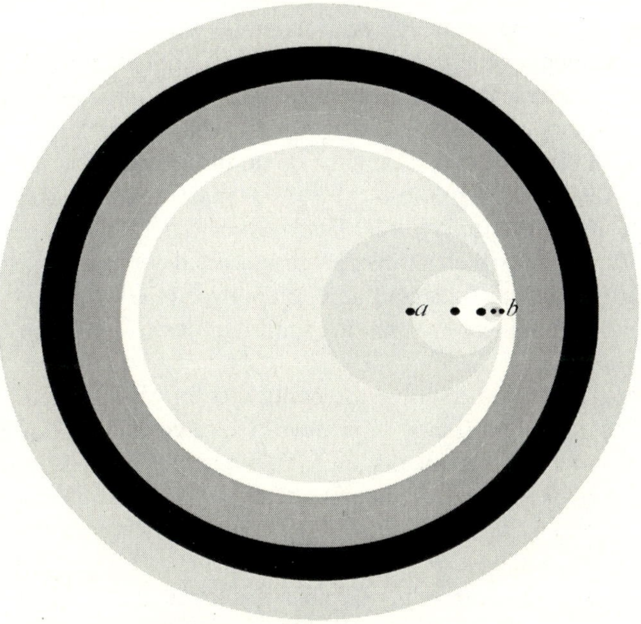

Figure 1.2. Artistic expression of motion from *a* to *b* realized by Setareh Korkchi.

Chapter One

Why Motion from *a* to *b* Is Impossi

The Analytic Principles of Thought Declare Motion Self-Contradictory

According to the reports of Plato and Aristotle, the Greek philosopher Zen who was a disciple of Parmenides of Elea, wrote an astonishing treatise which he attacked the thesis that there is motion in the physical world by deriving contradictions from it. Among the analytic arguments Zeno designed to show that motion involving contradictions is impossible, we will concentrate on the most famous, the *argument of dichotomy.*

In the eighth book of *Physics*, Aristotle presents two formulations of Zeno's argument of dichotomy.[1] The first, which is the most familiar, states: If we are to traverse the Euclidean distance *ab* of 1 unit, we must first traverse half the distance, or $1/2^1$; and then half of what remains, or $1/2^2$; and again half of what remains, or $1/2^3$; and so on ad infinitum. Thus, an infinite series of halves ($a_1 = 1/2$; $a_2 = 3/4$; $a_3 = 7/8$; . . .) must be traversed successively if we are to reach the limit $b = 1$. But if an infinite series is a series without limit, how is it possible to traverse an unlimited series and reach the limit $b = 1$? Don't we have here a contradiction between the unlimited series and the imperative requirement to reach the limit that destroys motion along the unit distance *ab*?

Aristotle attempted to solve the contradiction of unlimited/limited by assuming that the unit distance *ab* is *actually* limited and *potentially* unlimited.[2] Thus, when we traverse the distance *ab*, we *actually* and by *essence* traverse with one step or with a finite number of steps the whole distance *ab* and only *potentially* and by *accident* traverse with an infinite number of steps the unlimited series of halves ($0 + 1/2^1 + 1/2^2 + 1/2^3 + \ldots$) (Fig. 1.1.).

In this way, motion from $a = 0$ to $b = 1$ is saved at the cost of excluding the cumbersome infinite by assigning to it a potential or accidental ontological status. This solution is incomplete, as it solves the problem of motion under only the particular condition that the infinite is *uniquely* potential and not actual.

We would like, however, to present a universal solution to the problem of motion in which motion takes place under all conditions independently of whether the infinite series of halves is potential or actual. What happens, in fact, if the above particular condition changes and the infinite series of halves is an actual infinite? Can we traverse *actually* an unlimited series and reach *actually* its limit? At this point, our initial contradiction of unlimited/limit reappears in a vicious circle.

THE ANALYTIC PRINCIPLE OF INEQUALITY AND TEMPORAL ORDER DESTROYS MOTION FROM *A* TO *B*

Let us try to traverse *actually ab*'s unlimited series by adding successively its infinite number of halves and see whether simple arithmetic can assure us the passage to the limit. In the first $1/2^1$ instant, we obtain the partial sum $a_1 = 0 + 1/2 = 1/2^1$; in the next $1/2^2$ instant, we obtain the partial sum $a_2 = 0 + 1/2^1 + 1/2^2 = 3/4$; in the next $1/2^3$ instant, we obtain the partial sum $a_3 = 0 + 1/2^1 + 1/2^2 + 1/2^3 = 7/8$; and so on forever.

What have we obtained from this successive addition of an infinite number of halves? The answer is: merely an infinite sequence of partial sums: $a_1 = 1/2$; $a_2 = 3/4$; $a_3 = 7/8$; . . . $a_n = 2^n - 1/2^n$; and so on. The limit and total sum *b* is nowhere. To put it another way, *b* is somewhere but is inaccessible to the indefinitely approaching a_n. Indeed, no matter how close to its limit the converging sequence of partial sums a_n is, a_n must always be infinitesimally before and less than its limit and total sum *b* if the analytic principle of inequality and temporal order is to be satisfied: $a_n < b$.[3] This means that simple arithmetic employing temporal order can in no way realize the passage from the unlimited series a_n to the limit *b*. In fact, the best thing we can do with simple arithmetic is to compute up to the partial sum a_n and then leaving the total sum *b* incomputable. It follows that insofar as we perceive through our finite cognitive faculties of particular sensibility and analytic understanding the multiplicity of things a_n and *b* as contradictory points existing successively on a Euclidean line, continuous motion between consecutive points such as the computation of the total sum *b* by the successive addition of its parts a_n is an impossible task.

ANALYTIC CONVERGENCE AND THE APPARENT PASSAGE TO THE LIMIT

Because inequality and succession between the contradictory points a_n and *b* of the Euclidean line destroy continuous motion from a_n to *b*, the main question posed by the modern analytic theory of infinite convergent series is how to realize the passage to the limit despite this difficulty and without refuting the analytic principles of organization of our Euclidean observable world. This is accomplished by reducing inequality and succession into something infinitely small but not minimum. Indeed, any reduction of their inequality and succession into zero means that unequal things are equal, thereby violating the analytic principle of inequality and temporal order that governs the Euclidean observable world. The reduction of the inequality between a_n and b into something infinitely small enables us to count it *as if* it is zero and hence *as if* it is an equality without committing an appreciable error. This apparent equality between a_n and b enables us in turn to replace the variable a_n by its equal and constant b, and therefore to effect the passage to the limit b.

Thus, since the time of Cauchy, the modern Aristotelian mathematician defines motion from the unlimited series a_n to the limit b by the following formula:

$$\lim_{n \to \infty} a_n = b,$$

where n is a variable, finite number that increases without limit and ∞ is a constant, infinite number taken as the inaccessible maximum limit of the indefinitely increasing n. This formula gives an analytic definition of the infinite convergent series. The series a_n converges to its maximum limit b without ever reaching b if, and only if, the infinitely small difference $b - a_n = 1/n$ converges to its minimum limit $1/\infty = 0$ without ever reaching 0 as n of a_n, by passing through the infinite succession of finite numbers, increases indefinitely without ever reaching its maximum limit ∞. We call this unlimited series a_n deprived of its maximum limit b *indefinite* or *potential infinite* (Aristotle). It is a Euclidean *relative infinite* verifying analytic principles of organization such as the principle of contradiction, which stipulates that nothing unlimited is limited.

The infinitely small inequality between the consecutive terms $a_n < b$ enables us to behave *as if* there is an equality and continuous motion between them. In reality, however, there is a persistent discontinuous inequality that, no matter how small, destroys continuous motion. To put it another way, this infinitely small inequality is simultaneously an infinite inequality generating an infinite separation and distance between a_n and b. Indeed, insofar as the difference $\infty - n$ is always equal to ∞, the distance between n and ∞ is always ∞: that is, infinite or maximum. No matter how great n is, n is as far off from ∞ as the least finite number; hence, n is always the least part of ∞.

Let us establish a 1 to 1 correspondence between the unlimited sequence of integers n whose limit is ∞ and the infinite sequence of partial sums a_n whose limit is b:

1	2	n	∞
a_1	a_2	a_n	b

If we have $\infty - n = \infty$ and $b - a_n = 1/n$, and if by virtue of the assumed 1 to 1 correspondence we have

$$n = a_n \text{ and } \infty = b,$$

then if we replace in the difference $\infty - n$, ∞ by b, and n by a_n, we obtain

$$b - a_n = \infty.$$

We have then:

$$b - a_n = 1/n \text{ and } b - a_n = \infty,$$

which shows that the infinitely small difference and distance between a_n and b is infinite. The ancient problem of motion is posed again: How can we traverse the infinite distance that has no limit and attain the limit b?

To sum up, no matter how much the increasing a_n approaches b, a_n is both at an infinitely small distance from b and at an infinite distance from b similar to the least finite number a_0 of the series of finite numbers a_n. Not only is the indefinitely approaching a_n incapable of reaching its end and last number b at point b, but it is also incapable of moving beyond itself insofar as we regard it as similar to the least finite number a_0 at point a. Continuous motion along the Euclidean unit distance ab is impossible because the indefinitely varying a_n can neither begin nor end its motion. Such is the paradox of motion in which the impossibilities to begin and to end motion are different aspects of impossible motion in the immobile Euclidean analytic space.

The analytic theory of the infinite convergent series has failed to deliver the promised solution to the problem of motion because it developed according to our finite individual mode of perceiving the physical world's multiplicity of things successively in conformity with the analytic principle of inequality and temporal order. Because inequality and succession between different things, for instance $a_n < b$, interrupt continuous motion, the aim of the analytic theory of the infinite convergent series is to allow the discontinuous inequality to become infinitely small so that it can be counted *as if* it were a continuous equality without committing an appreciable error. This attempt has proved unsuccessful, however, as the desired infinitely small separation and distance between a_n and b can in no way neutralize the *actually* infinite separation and distance between n and ∞ and therefore the actually infinite separation between a_n and b. The apparent equality based on the infinitely small difference and inequality between a_n and b disappears to reveal the naked truth: the existence of an actually infinite gulf between a_n and b that no analytic theory of the infinite convergent series can bridge.

What the analytic theory of the infinite convergent series accomplishes is not the solution to the problem of motion; it is merely its ephemeral *regulation* by dissimulating the impossibility of solving it by way of analytic means. In fact, because our finite retinal cells at rest detect uniquely the finite part of light traveling empty space at the finite speed c, we are incapable of perceiving the multiplicity of things $a_n \neq b$ simultaneously as a unitary whole; instead, we perceive them successively as contradictory and isolated individuals deprived of unity and separated by an infinite distance. This shows us that our finite analytic experience, far from being the solution to the problem of motion, is its very origin!

It follows that solving the problem of motion resides outside the analytic principles of our Euclidean time-conditioned observable world generated by our finite individual senses and their corresponding cognitive faculties of particular sensibility and analytic understanding and inside their negations.

NOTES

1. Aristotle, *Physics*, trans. Henri Carteron (Paris: Les Belles Lettres, 1986), VIII (8), 263 a_4- a_{11}.

2. Ibid., 263 b_3–b_9.

3. The analytic principle of inequality and temporal order states that if any two things a and b are different, then they are unequal and consecutive. Thus: If $a \neq b$, then either $a < b$ or $b < a$.

Chapter Two

The Founding Principle of Continuous Motion

How can the infinite converging series a_n reach and coincide with its maximum limit $b = 1$ if it is separated by an actually infinite distance? Only if the Euclidean infinite distance between a_n and b has simultaneously a zero magnitude and hence is simultaneously an extensionless point can we coincide a_n with b and therefore realize the continuous passage to b. This means the *infinite* difference separating n from ∞ must be simultaneously a *zero* difference, as in:

$$(\infty - n = \infty)(\infty - n = 0) \text{ or } \infty - n = \infty \times 0,$$

(where the multiplication or conjunction sign $\times$ designates both) in order to equalize n with ∞, such as $n = \infty$; $1 - a_n = 1/n$ with 0, such as $1/n = 0$; and a_n with 1, such as $a_n = 1$.

Mathematicians call *limiting point b*, which they take as the real point, the infinite distance's outer point that determines its contained inner infinitely many points. Influenced by Aristotle's concept of limit, the philosophically oriented mathematicians consider the real limiting point b i) as the first and last individual point originating and ending the unlimited series of points beyond which it is not possible to find any point; ii) as the infinite totality of points; and iii) as the unifying principle uniting infinitely many points.[1]

Cosmologists call the limiting point b *cosmic* or *space-time singularity* of infinite curvature, density, and energy, which they identify with the outer part of the universe's extreme boundary, also called the limiting boundary. Considered as the first point originating everything, the cosmic singularity is called the *Big Bang*, whereas considered as the last point ending everything, the cosmic singularity is called the *Big Crunch.*

The ancient metaphysicians (Ionian Greeks, Persians, Hindus) called the limiting point *b* the *point-fire* and identified it with the cosmic infinite mind, intellect, nous (νοῦς), logos (λόγος), or soul (ψυχή, *anima*) that permanently animates and unifies the infinitely extended and divided universe. In this sense, the infinite and self-ruled mind (Anaxagoras) is a *mental* source, an *energy* source, and a *life* source having the power of self-consciousness, self-motion, and self-order (spontaneous and reflexive order). Being a fundamental constituent of matter, the infinite and self-ruled mind assigns consciousness, kinetic energy, life and form to the universe's unconscious, inert, lifeless and formless matter. The infinite and self-ruled mind works also as a unifying principle, which we may logically call the universal equivalence principle and poetically call the principle of universal love or sympathy, which realizes the unity of opposites regardless of their difference and distance. It is through this unifying point that the universe holds together regardless of its infinite extension and division, and that distant parts of the universe immediately communicate with one another regardless of their infinite distance.

Modern metaphysicians call the limiting point *b* the intellectual and intelligible *real thing-in-it-self* (τό κατ'αὐτό) that grounds the Euclidean series of sensible (observable) parts a_n called appearances or phenomena. The real thing *b* may be i) the *point-matter* of the phenomenon (Kant); ii) the real physical body defined as the sum total of its infinite series of observable parts a_n (Russell); if the limit of the physical body is its surface, then the surface is the limit of its being and of the knowledge of its being; iii) the ultimate constituent of the physical body, called an *element*, and residing in the deepest level of the physical body; iv) the *real self*, or the *infinite mind* defined as the sum total of the brain's infinite number of observable neurobiological effects.

Theologians call the limiting point *b* the eternal Supreme Being—the Being of all beings or the *ens realissimum* possessing maximum reality. They also assimilate the limiting point *b* with the Demiurge, which they conceive as a self-caused and self-moved being (τό κατ'αὐτό κινούμενον ὂν) endowed with the power to be both the cause and the effect of its proper being and motion as well as with the power to be both the first and final cause of its contained Euclidean series of sensible parts a_n. Through its infinite curvature, the limiting point *b* tells the incomplete sensible part a_n *where* to move, *why* to move, and *how* it is possible to move to its maximum limit *b* in order to realize its completion.

Taking this into consideration, we replace Cauchy's analytic formula with Cantor's synthetic formula in order to define the infinite convergent series a_n, which effectively realizes the continuous passage to the maximum limit *b*:

$$\lim_{n=\infty} a_n = b.$$[2]

The infinite series a_n converges to b and actually reaches its maximum limit b if, and only if, n of a_n travels at the same time through the Euclidean infinite distance (or infinite series of sub-distances) separating n from ∞ and beyond the Euclidean infinite distance through its outer limiting point b of zero distance that unites n with ∞. This synthetic unity geometrically realized by the unifying limiting point b of infinite curvature is logically expressed by the equality and simultaneity link between n and ∞; it enables us to assert that n moves instantaneously to the limit ∞ if, and only if, n is equal and simultaneous to ∞ and ∞ is equal and simultaneous to n. Similarly, a_n moves instantaneously to b if, and only if a_n, by virtue of its equality link with b, is equal and simultaneous to b and b is equal and simultaneous to a_n.

The principle that unites the unlimited series n, $1/n$, and a_n with their respective maximum limits ∞, 0, and 1 is the finite-infinite equivalence principle, which stipulates the equality of the unlimited series with its limit, that everything is both unlimited and limited, that everything unlimited is limited, and that everything limited is unlimited without contradiction or paradox.

The finite-infinite equivalence originates historically with the Pythagoreans (sixth century BCE) and Plato (fifth to fourth century BCE), who affirmed that every being numbered by 1 is governed by two principles, the finite (πέρας) and the infinite (ἄπειρον);[3] it follows that the *whole one* is in reality an *infinite whole one*, both one and infinitely many. The Pythagoreans situated the finite at the right column and the infinite at the left column and attributed to them ten properties: for example, they attributed the odd number 1, which is a male number, to the finite and the even number 2, which is a female number, to the infinite. The synthetic unity and simultaneity of the finite and the infinite generates the finite-infinite equivalence principle; it also generates the number 1 computing the total sum of infinitely many parts. Therefore, 1 is both a number and a principle of unity. Within 1, the number and the principle are reconciled. The intuitive idea that the being numbered by 1 is a synthesis of the finite and the infinite was already circulating among the natural philosophers of Ionia and India during the seventh and sixth centuries BCE.

The finite-infinite equivalence principle is geometrically realized by the infinite distance's outer limiting point b. By virtue of immediately connecting the unlimited series of parts with its respective maximum limit, the unifying limiting point b allows continuous motion between them. We call *transcendental* limiting point b the hypersensible or intelligible whole b, which is simultaneously a sensible part a_n belonging to the Euclidean series of sensible parts a_n. Because the intelligible whole b, regarded as the condition of the series of sensible parts a_n is at the same time a sensible part a_n, the series is *mathematical* (Kant) and the intelligible whole b is empirically accessible because it is sensible.[4] The intelligible whole, which is simultaneously a sensible part, reminds us of the Platonic ideas, which though separated from sensible things are immanent to sensible things. This state of being both intelligible separated from sensible things and sensible not-separated from sensible things is possible without absurdity because the Platonic ideas are not simple individuals but rather complex wholes or principles unifying opposites such as the intelligible with the sensible.

The *transcendental* limiting point *b*, which is an intelligible whole regarded at the same time as a sensible part and hence as a self-contained point, must be distinguished from the *transcendent* limiting point *b*, which is a purely intelligible whole not belonging to the Euclidean series of sensible parts a_n. Because the condition and maximum limit of the series of sensible parts a_n is a uniquely intelligible whole *b*, occurring either before or after the series of sensible parts such as $b < a_n + a_n < b$, the series is *dynamic* (Kant) and the uniquely intelligible whole *b* is empirically inaccessible because it is not a sensible part.

We have shown in chapter 1 how this dynamic series of parts involving a discontinuous inequality and temporal order between its parts—namely, between the sensible part a_n and the intelligible whole *b*—destroys continuous motion from a_n to *b*. It follows that only a mathematical series involving a real equivalence between the series of sensible parts a_n and its intelligible whole *b* can allow and ground continuous motion from a_n to *b*.

At this point, the Anaxagorean theory of motion enters our discussion. Because it is unanimously agreed by the natural philosophers of Ionia (sixth to fifth centuries BCE) that motion between contradictory opposites is impossible, it follows that in order to generate one opposite from another, we must assume the presence of one opposite in the other.[5] In this way, opposites generate each other insofar as they are already present in each other and thus are *contraries*—that is, equal opposites existing simultaneously since eternity or infinite time—and not ephemeral *contradictories* existing successively since a finite time.

Using the same Anaxagorean solution to our problem of motion, we assert the following: In order to generate the maximum limit $b = 1$ from something infinitely different, such as a_n, we must assume that at the deepest recesses of physical reality, which are actually imperceptible by our finite individual senses, the maximum limit $b = 1$ is already present in the unlimited series a_n and that the unlimited series a_n is already present in $b = 1$. The same holds true for the remaining unlimited series *n* and $1/n$ and their respective maximum limits ∞ and 0.

Now, if everything—for example, a_n and *b*—is already in everything—that is, in both a_n and *b*—then the experienced successive motion from a_n to *b*, involving a change of position at different instants with an indefinitely increasing finite speed, is an appearance or a perceptual illusion that takes the appearance of succession *as if* it were the very immanent property of the real physical world. It follows that real, continuous motion independent of our analytic particular perception of things ultimately must be at the deepest and highest level of reality instantaneous (immediate), involving a change of position from a_n to b in one instant and with an infinite speed.

Thus, for the natural philosopher Anaxagoras, mutual generation between elements is an appearance, a delusion. In fact, for each physical body, generation is simply a process of extracting whatever already exists in the physical body considered as a universal receptacle (τό πανδεχές) or whole containing everything at all times. This Anaxagorean thesis leads us to conclude that generation does not produce change. Because the physical body is a synthesis or mixture of everything existing and being fixed within the physical body since eternity, we conclude that generation and destruction, which are successive phases of linear time, do not really exist in themselves independently of particular or individual observation; in this sense, they are appearances or accidents of our finite individual senses: for example, of our finite retinal cells at rest, which detect uniquely the finite part of light's speed traveling empty space at the finite speed c and finite frequency f. The finite speed c of light creates the illusion of time delay and temporal order between here and there, now and then, near and distant.

It follows, then, that *real* change must be spatial (non-temporal) assimilated to ideal locomotion, regarded as primary motion and which we define as change of place within an instant, and at infinite speed. Because the instantaneous occupation of different places separated by any distance is motion within the same place (which we call rotation), it follows that the rotating sphere is the best figure for reconciling motion (difference of place) with constancy (sameness of place) and therefore for realizing the permanence of motion. In fact, the sphere is the geometric form of the universe's physical body permanently rotating at an infinite speed under the constant action and rule of the infinite mind (νοῦς), located at the outer extreme boundary of the universe.[6]

We have called *relative* or *potential infinite* the Euclidean infinite series of parts deprived of a limiting point for its completion (see chapter 1). Regarded as an incomplete part having everything left outside, the potential infinite is in reality a variable that increases or decreases indefinitely—that is, without limit—but always remains finite. It is therefore an *apparent* or *improper infinite*.[7] Because it generates the tension, frustration and absurdity of infinite regression, we can also call it *evil infinite*. When $n < \infty$, n is an incomplete potential infinite deprived of its limiting point ∞ and hence ruled by the analytic principle of inequality and temporal order, which is the organizing principle of our Euclidean observable world.

On the other hand, we call *actual infinite* or *infinite in act* the Euclidean infinite series of parts with a limiting point that completes and actualizes the infinite series. In fact, by assigning a fixed position to the potential infinite series without destroying its unlimited nature, the limiting point stabilizes and therefore actualizes the infinite series. Infinity ceases to be potential, imaginary, illusory, or utopic and becomes actual and real; it ceases to be nowhere and acquires a topos, which is the very limiting point containing the infinite series.

Because the limiting point unifies the infinite series into one thing, it transforms the infinite series from an *indefinite part* lacking everything into an *infinite whole* containing everything and lacking nothing. In short, the limiting point transforms the Euclidean *relative infinite* into a non-Euclidean *absolute infinite*—that is, into a maximum incapable of further variation. It is the *real* and *proper infinite*—an infinite with a limiting form governed by the synthetic finite-infinite equivalence principle and represented geometrically by the infinite sphere. For instance, when $n = \infty$, n is a complete infinite series with the limiting point ∞ behaving as a spherical whole that contains the infinite series n at once and allows this infinite series to behave as one infinite whole in act. The limiting point ∞ is also, by virtue of the equivalence $n = \infty$, both the containing spherical whole ∞ and the contained Euclidean part n without implying a contradiction, and in this sense it is free from the Euclidean analytic principles of organization. For example, the infinite whole ∞ refutes the Euclidean principle according to which the whole is greater than any of its parts and the analytic principle of inequality and temporal order, which asserts that if any two things are different, then they are unequal and have a linear temporal order between them.

As a matter of fact, the infinite whole satisfies its proper non-Euclidean and synthetic principle of organization, according to which a whole is infinite if it is equal to any of its proper parts. We then take this infinite whole or absolute infinite reconciling opposites as the immanent body of the real physical universe

NOTES

1. According to Aristotle, *limit* is: i) the extremity of a thing, that is, the first point beyond which it is not possible to find anything and the first point within which all the points are; ii) the form of what has magnitude, which is the body; iii) the end of anything, that to which, not from which, a movement or action proceeds, but sometimes both origin or principle (ἀρχή) and end; iv) the what (*quiddité)*of anything, for this is the limit of knowledge, but also the limit of the known thing. See also Aristotle, *Metaphysics*, trans. J.Tricot (Paris: Librairie Philosophique J.Vrin, 1984), Δ, 17.

2. Bertrand Russell, *The Principles of Mathematics* (London: George Allen & Unwin, 1979), Chapter XXXIV. Limits and Irrational Numbers, 285.

3. Aristotle, *Metaphysics*, trans. J.Tricot (Paris: Librairie Philosophique J.Vrin, 1984), A, 5, 986a 20-30.

4. For the distinction between mathematical and dynamical series, see Kant's *Critique of Pure Reason* (London: Everyman's Library, 1991), Transcendental Dialectic, Book II, Chapter II. The Antinomy of Pure Reason, Section IX, II Concluding Remark on the Solution of the Transcendental Mathematical Ideas and Introductory to the Solution of the Dynamical Ideas, 314-316

5. For a presentation of the Anaxagorean theory of the being see Aristotle, *Physics*, trans. H. Carteron (Paris: Les Belles Lettres, 1983), I (4) 187*a–b.*

6. Concerning the Anaxagorian infinite mind (νοῦς) see also G.S. Kirk, J. E. Raven, and M. Schofield, *The Presocratic Philosophers* (Cambridge: Cambridge University Press, 1988), 352-384.

7. The nineteenth-century mathematician Cantor called *improper infinite* the incomplete infinite series, which is not closed and determined by a limiting point. Per contra, Cantor called *proper infinite* the infinite series closed and determined by a limiting point. See "*Fondements d'une théorie générale des ensembles*", which appeared in 1883 in the *Mathematische Annalen* XXI and in the *Cahier pour l'Analyse:* "*La Formalisation*" (Paris: Editions du Seuil, 1969), 35-36.

Chapter Three

Geometric Solution to the Problem of Motion

The Infinite Sphere

Thus if there are many things, it is necessary that they are both small and large; so small as not to have magnitude, so large as to be unlimited.

— Zeno

Let us represent the whole of the infinite series—that is, the infinite whole—by an infinite sphere or an infinite circle, which is the great circle of the infinite sphere, centered on us at the origin *a* and having an infinite radius. Following Plato, who considers the infinite as a complex whole and divides it into contrary (equal and opposite) parts, such as the great infinite (τό μέγα ἄπειρον) and the small infinite (τό μικρόν ἄπειρον), we divide the infinite sphere into i) the inner infinitely great or greatest sphere of infinite radius, and ii) into the outer infinitely small or smallest sphere of zero radius.[1]

If the curvature of the surface of the infinite sphere is the inverse of the radius, then the outer infinitely small sphere of zero radius has a surface whose curvature is infinite, namely a convex point, whereas the inner infinitely great sphere of infinite radius has a surface whose curvature is zero, or negative infinite; that is, a concave point. The outer convex point—called the limiting point or cosmic singularity—of the infinite sphere circumscribes the inner flat or concave part of the infinite sphere without closing it. It is necessary that the sphere's infinite radius is partially delimited by its inverse zero radius in order to assign a finite radius 1 to the sphere, which is a figure limited by construction. In fact, if the radius of the infinite sphere was *uniquely* infinite, then it would not be a radius and the infinite sphere would be a geometric and a logical impossibility, as there would be a contradiction between infinity and the sphere's inherently limited unit radius. Only if we divide the radius and the surface of the sphere into contrary (equal and opposite) parts in conformity with the Platonic division of infinity into great and small can we reconcile the finite magnitude and the infinite magnitude within the same sphere.

The limiting point assigns to the infinite sphere the quality of the greatest sphere having nothing greater than itself, as well the quality of the smallest sphere having nothing smaller than itself. Thus, this limiting point of zero radius, which is a compact nothing of infinite energy density, is both the origin and the end of all magnitudes, according to extension and division. As the origin of all magnitudes, nothing is the smallest and lowest magnitude that cannot be further divided; it is therefore contained in infinity. As the end of all magnitudes, nothing is the greatest and highest magnitude that cannot be further extended; it therefore contains infinity. Ultimately, the limiting point is both a containing whole containing all magnitudes and a contained part contained in all magnitudes—that is, a self-contained being, which similar to the living being verifies the synthetic principle of reflexive order (self-order, spontaneous or autonomous order).

Let us consider the infinite sequence a_n and its limit b as points of the Euclidean radius ab of the infinite sphere such that the origin point a_0 of the sequence a_n is at the center a of the infinite sphere, the infinite sequence a_n is somewhere between a and b, and the end point b is on the infinite sphere, that is, on the limiting boundary b of the infinite sphere. If the radius ab is Euclidean and if Euclidean space is a set of points that are at unequal distance from a chosen common origin $a_0 = a$, then insofar as a_n and b are points of the Euclidean radius ab, they are at unequal distance from their common origin a and are therefore unequal, such as: $aa_n < ab \rightarrow a_n < b$. As unequal points of the Euclidean radius ab, they verify the analytic principle of inequality and temporal order, according to which different points, say $a_n \neq b$, separated by an infinite distance are experienced by our finite individual senses as unequal and time-conditioned, such that a_n is less than and before b: $a_n < b$. It follows that continuous motion on the Euclidean radius ab, where different points $a_n \neq b$ separated by an infinite distance are experienced as unequal time-conditioned points, is impossible.

Because continuous motion from a_n to b requires that a_n and b coincide, that they are equal and therefore equidistant from their common origin $a_0 = a$, we must situate them on the limiting boundary b of the infinite sphere in order to realize motion. Indeed, if the sphere is a set of points at an equal distance from a chosen common origin $a_0 = a$, then insofar as a_n and b are points on the infinite sphere, they are equidistant from their common origin a_0 at the center a and are therefore equal, such as: $aa_n = ab \rightarrow a_n = b$ (Fig. 3.1).

Now, as equal points on the limiting boundary of the infinite sphere, they verify the synthetic principle of equivalence and zero temporal order,

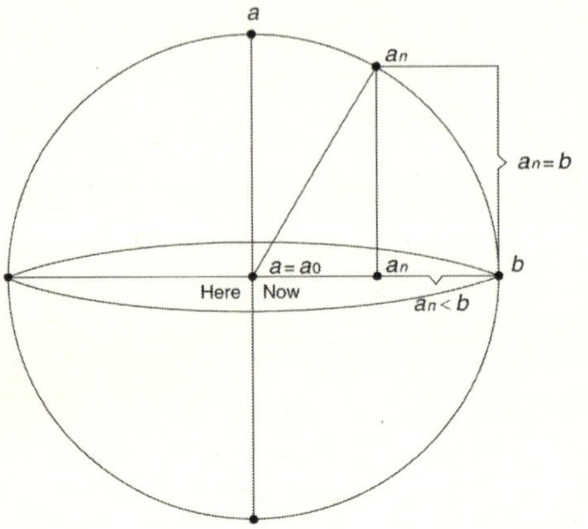

Figure 3.1. Here the infinite sphere is defined as the common boundary (interface) between the infinitely great sphere of infinite radius and the infinitely small sphere of zero radius, called the limiting point *b* or cosmic singularity. Real continuous motion is not along the Euclidean radius *ab* where the distinct points $a_n \neq b$ are unequal and discontinuous ($a_n < b$), but rather along the non-Euclidean radius *ab* identified as the arc *ab* on the infinite sphere, where the same distinct points $a_n \neq b$ are equal and continuous: ($a_n = b$).

Figure 3.2. Artistic expression of the infinite sphere showing the limiting boundary containing the infinitely many. Realization: Setareh Korkchi.

according to which different and distant points $a_n \neq b$ experienced by our individual senses as unequal in the Euclidean radius *ab* are thought by our faculty of synthetic reason as equal $a_n = b$, that is, as having a zero distance between them on the non-Euclidean limiting boundary of the infinite sphere. It follows, then, that continuous motion from a_n, taken as the starting point *a*, to the end point *b* is instantaneous despite their infinite separation and that the continuous rotation of the infinite sphere is not an impossibility—as asserted by Aristotle in his treatise *On the Heavens*—but rather a necessity insofar as we divide the distance between the infinite sphere's radii into an inner distance of infinite magnitude and into an outer distance of zero magnitude.[2]

Thus, the starting point *a* on the rotating infinite sphere can move across the infinite distance separating *a* from *b* and immediately reach the end point *b* with one rotation in a free, effortless, and timeless manner precisely because the infinite distance that separates them is simultaneously the zero distance that unites them. From the composition of these orthogonal distances—namely i) the inner infinite distance seen horizontally along the Euclidean line of sight *E* and ii) the outer zero distance seen vertically along the non-Euclidean line of sight E', also called the mental line, we derive the complex and indeterminate unit distance *ab*, which we define as the product or ratio of contrary (equal and opposite) distances having infinite and zero magnitudes:

$$(ab = \infty)_E (ab = 0)_{E'} \qquad \text{or} \qquad ab = \infty \times 0 = 1 \qquad \text{or} \qquad ab = \infty/0 = 1.$$

The complex unit distance *ab* satisfying any of the above formulas and assimilated to the arc *ab* of the infinite sphere's great circle is not a paradox: It is the very solution to the paradox! It is grounded in the synthetic finite-infinite equivalence principle, according to which everything is both finite and infinite (stipulated in chapter 2) and expresses what the ancient Greeks had thought twenty-five hundred years ago as seen in Zeno's argument about magnitude: that if anything exists, it must have a magnitude that is both small and large, "so small as not to have magnitude, so large as to be unlimited."[3] The composition of zero magnitude and infinite magnitude assigns to the existing thing the finite magnitude 1, which is both a real number and the synthetic principle of the unity of opposite things (distances) or parts (magnitudes) of the same thing (distance). Thus, the Zenonian unit distance *ab* satisfies the non-Euclidean postulate according to which there are many magnitudes that measure the same unit distance *ab*, or there are many magnitudes that measure many distances between the points *a* and *b*. This impossibility of assigning a unique magnitude to the non-Euclidean unit distance *ab* reveals the indeterminate and complex nature of its structure, which is neither finite nor infinite and yet is all of this simultaneously. By mutually negating the opposite determinations finite/infinite, we obtain a non-Euclidean unit distance *ab* that is free of the partiality of determination and hence indeterminate and impartial by comprising the totality of determinations.

The indeterminate and complex nature of the unit distance *ab* is also expressed by the word *di-mension*, whose ancient Greek root, δίς-μετρον, means twice measured. If we consider the length of the unit distance *ab* as *one dimension*, then in conformity with its Greek root, it must have two measures: for example, following Zeno's line of thinking, it must have an infinite dimension and a zero dimension. This means that any single dimension in the physical universe is a complex whole situated on the synthetic diagonal axis that can be resolved into an infinite dimension along the horizontal axis of the Euclidean line of sight *E* and into a zero dimension along the vertical axis of the non-Euclidean line of sight *E′*. In this way, the one dimension divisible into two dimensions reflects the complex structure of our physical universe and our synthetic reason, defined as the faculty of conceiving this complex structure.

THE UNIVERSAL MOTION OF THE INFINITE SPHERE

Heaven attains its end immediately with one movement.

—Aristotle, *On the Heavens*

Having shown that continuous motion is on the limiting boundary of the continuous infinite sphere, we conclude that any point a_n (for instance a galaxy or a universe of galaxies) appearing to our individual senses as indefinitely accelerating away from us at the center *a* and toward its limit *b* on the Euclidean radius *ab* is in reality on the non-Euclidean radius (arc) *ab* participating in the continuous rotation of the infinite sphere at a maximum speed. Because on the limiting boundary of the infinite sphere the infinitely distant points a_n and *b* are equal—that is, they have a zero distance between them—they coincide, and motion between them is instantaneous at infinite and zero speeds; the product of these two infinite speeds is, according to the Zenonian conception of magnitude, the real speed 1. In turn, we regard this complex unit speed composed of contrary infinites as the real speed of maximum motion, which is an instantaneous motion at a distance.

Thus, the infinite sphere, by virtue of rotating at infinite and zero speeds whose product is the real speed 1, has the power to communicate universally with all its infinitely distant and equal points at once. This instantaneous communication at an infinite distance constitutes the infinite sphere's immanent and absolute action, which is simultaneously the universal action of its universal laws. In fact, if there was no instantaneous communication among the infinite sphere's infinitely distant points, how could its infinitely distant points obey the *same* laws, called universal laws, and have the *same* properties, for example the same temperature, density, and time regardless of their distance and difference?

It is clear that the perfect uniformity of the infinite sphere (where all its points and directions are equal and have the same properties, which we call universal and common properties) and the perfect universality of its laws require that communication among its infinitely distant points be instantaneous and that this instantaneous communication allow them to function as a continuous and unified infinite whole governed by the synthetic equivalence principle.

Now, the immanent, absolute, and universal action of the rotating infinite sphere is the action of immobility that conjoins through its complex unit speed maximum motion of infinite speed with minimum motion or rest of zero speed. In this sense, the action of immobility, which the ancient thinkers called the activity of immobility (ἐνέργεια ἀκινησίας), is the power of exerting the greatest action while being at rest: that is, with least action.[4]

As a rotating agent imparting maximum and universal motion to itself without being moved, the infinite sphere is an effortless actor in continuous self-motion. As the first and final limiting point *b* circumscribing, unifying, originating, and ending its inner unlimited sequence of parts a_n, the infinite sphere is a self-limited, self-unified, self-originating, and self-ending body verifying the synthetic principle of reflexive causality (self-causality), which is an alternative expression of the synthetic principle of reflexive - order (self-order, free or spontaneous order).

According to reflexive causality, everything that happens is a cause of itself (*causa sui*). Because the infinite sphere is both a cause and an end of itself, its causal action from the cause to the end is nonlinear (circular) and universal, traveling across and beyond the sphere's inner infinite distance instantaneously with infinite and zero speeds. The infinite sphere thus attains its cause (origin) and end immediately with one rotation regardless of its inner infinite size. In fact, as we have argued, this infinite size balanced by the infinite sphere's outer zero size has no contradictory and destructive effect on the infinite sphere's rotation.

Self-causality (autonomous or spontaneous causality) is the founding principle of the self-organizing and self-moving living being. It is also the founding principle of the infinite sphere, which is the geometric figure of the infinite whole or infinite body. It follows that the infinite body is the body of the spontaneously rotating and vibrating living being whose different manifestations are the infinite Universe, God, and Mind.

THE LIMITING BOUNDARY OF THE INFINITE SPHERE

Only God is sufficiently intelligent and powerful
to unite many things into one thing
and divide one thing into many things.

—Plato, *Timaeus*

Before continuing with the investigation of the nature of the curvature of the infinite sphere, let us define the meaning of that curvature. If we draw circles of different radii, we will notice that the curvature K of the one-dimensional circle varies in inverse proportion to its radius R, in other words, the curvature is the reciprocal of the radius:

$$K = 1/R.$$

Moving from the one-dimensional circle to the two-dimensional surface of a sphere, the curvature K of the two-dimensional surface is the reciprocal of the sphere's squared radius: $K = 1/R^2$. Now, if R is extended to infinity under the action of a stretching force—for example, of repulsive gravity—we obtain an infinitely great sphere whose curvature is zero, and therefore its surface is flat:

$$R = \infty \rightarrow K = 1/R^2 = 1/\infty = 0.$$

This means the infinitely great sphere of infinite radius is equivalent to the Euclidean plane. If R shrinks to zero under the action of a shrinking force—for example, of attractive gravity—we obtain an infinitely small sphere whose curvature is infinite, and therefore its infinitely curved surface is a convex limiting point—that is, a cosmic singularity:

$$R = 0 \rightarrow K = 1/R^2 = 1/0 = \infty.$$

If we define the infinitely great sphere as the limit b of an infinite sequence of homocentric spheres of increasing finite radius caused by a stretching force, then at the limit b the extending sphere is infinitely stretched having an infinite radius and an infinitely extended and infinite- dimensional surface. Similarly, if we define the infinitely small sphere as the limit b of an infinite sequence of homocentric spheres of decreasing finite radius caused by a shrinking force—for example, by attractive gravity—then at the limit b the shrinking sphere is infinitely contracted, having a zero radius and an infinitely divided and zero-dimensional surface. The composition of both opposite spheres gives an infinite sphere with an infinitely extended and infinite-dimensional surface, which is embedded in an extensionless nothing of zero dimension, and that we regard as the greatest and highest dimension.[5]

If we perceive the infinite sphere from the inside and horizontally along the Euclidean line of sight *E*, we observe that the curvature of its limiting boundary is zero and that the limiting boundary itself is flat, that is, a Euclidean plane. We conclude, then, that the limiting boundary is nowhere, or, to put it in another way, that the limiting boundary is imaginary or utopic. Thus, according to the famous definition of Hermes Trismegistus, an infinite sphere is an intelligible sphere, the center of which is everywhere and the boundary of which is nowhere.

Hermes Tris-megistus (τρίς μέγιστος), which means thrice-greatest—for example the greatest philosopher, priest, and king—was a mythical figure of the Greek Egyptian antiquity (Hellenistic period) that emerged from the combination of the Greek god of communication, Hermes, and the Egyptian god of writing, Thoth. An ensemble of sacred texts encapsulating ancient wisdom revealed by *divine nous* (νοῦς) and called *Hermetica* was attributed to Hermes Trismegistus, a collective name that stands for many authors. The modern name for the Greek divine nous is our intellectual faculty of *infinite synthetic universal reason*, which we define here as the faculty of thinking the unity and wholeness of the infinite Euclidean space. The Hermetica discusses the nature of the cosmos, the mind, and the divine, and includes a method of personal ascension and emancipation from the constraints of finite existence.

Hermes Trismegistus used the idea of infinite sphere to geometrically define infinite God. Subsequently, the German philosopher and Renaissance Platonist Nicholas of Cusa (*De Docta Ignorantia*, fifteenth century CE) and the French philosopher and mathematician Blaise Pascal (*Pensées*, seventeenth century CE) used the hermetic definition of God to define the infinite Universe and Nature, which they assimilated to the infinite sphere.

If, however, the infinite sphere has no limiting boundary, then it is in reality an infinite Euclidean plane and cannot constitute a proper definition of God conceived as the being capable of thinking and sensing the unity, simultaneity, and limiting wholeness of the infinite plane. It follows that in order for the imaginary limiting boundary to become real, the Euclidean infinite plane must have an outside in the highest dimension that delimits it without destroying its inner infinity and flatness. This means that the real limiting boundary is self-divided into two coexisting parts—the inner flat surface and the outer infinitely curved convex surface. If we perceive the infinite sphere from the outside, we will observe the curvature of its limiting boundary to be infinite and the limiting boundary itself to be a real limiting point *b*—a compact nothing of infinite energy density assimilated to a cosmic singularity or a point-fire coextensive to life and mind (νοῦς). Because the real limiting point *b* or compact nothing is an immense point greater than any infinity, it contains the whole infinite Euclidean plane at once and attracts all directions of the Euclidean plane into one universal point working as the common constituent of all things as well as the principle of unity of all opposites. The limiting point *b* also allows its infinitely many things to behave as one whole thing; brings the infinitely distant things inside each thing and at a zero distance from each thing; and realizes the universal communication of the infinitely many and infinitely distant things within zero time and at a maximum speed.

By assigning a limit to the infinite Euclidean plane, the limiting point *b* transforms the Euclidean plane's imaginary limiting boundary existing nowhere into a real limiting boundary with a fixed universal position containing everything and existing everywhere.[6] It follows that the real infinite sphere is not the hermetic sphere the center of which is everywhere, its limiting boundary nowhere, but instead the sphere whose center and limiting boundary are everywhere (Figure 3.3).

Now, the curvature of this real infinite sphere, or, to put it in another way, of the real limiting boundary of the infinite universe, is the composition of the above zero and infinite curvatures that gives the positive, constant curvature

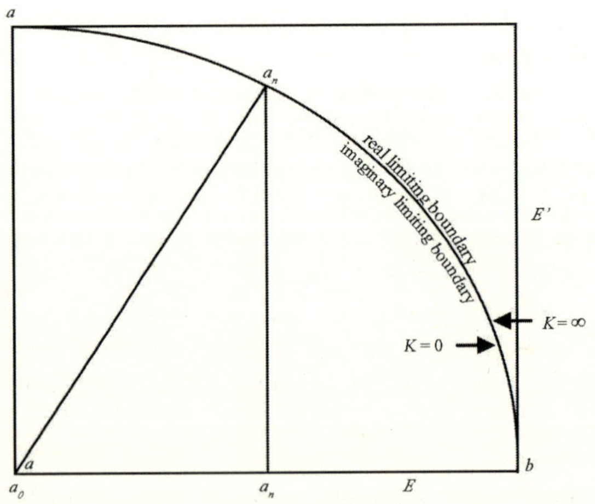

Figure 3.3. The real limiting boundary (surface) of the infinite universe is divided into the inner flat or concave part and the outer or convex part. The inner part is the Euclidean plane of zero curvature having no limiting boundary or an imaginary limiting boundary. An imaginary limiting boundary is a concave boundary infinitely curving inward, or away from the eye. The outer part is the limiting point *b* of infinite curvature, which is the real limiting boundary assigning position and reality to the imaginary limiting boundary. A real limiting boundary is a convex boundary infinitely curving outward, or toward the eye.

$K = 1$, which we define as the product or ratio of contrary curvatures having infinite and zero magnitudes:

$$(K = 0)_E(K = \infty)_{E'} \qquad \text{or} \qquad K = 0 \times \infty = 1 \qquad \text{or} \qquad K = 0/\infty = 1.$$

The positive curvature $K = 1$ of the real infinite sphere therefore is not simple and determinate having at one time a unique magnitude, namely the magnitude 1, but rather complex and indeterminate resolved into zero and infinite magnitudes. In a way, the finite positive and constant curvature $K = 1$ of the infinite sphere emerges as the intermediate and indeterminate factor, which is neither a zero curvature determining a flat surface nor and infinite curvature determining a limiting point and yet is all of this at once. In fact, we may think of the real limiting boundary of the infinite sphere defining the real physical universe as the complex place where something is something else, where one thing is divided into two things and two things are united into one thing, for example where the real limiting boundary of positive curvature equal to unity is divided into distinct limiting boundaries of zero and infinite curvatures and, conversely, the distinct limiting boundaries of zero and infinite curvatures are united into one common universal limiting boundary of positive curvature equal to indeterminate unity. Ultimately the real limiting boundary of the universe is both divided (τό διαιρετόν) and undivided (τό ἀδιαίρετον), both multiple and one.

The same holds true for the radius R of the real infinite sphere, regarded as the reciprocal of its curvature K: $R = 1/K$. Similar to Zeno's unit distance *ab*, the real infinite sphere of unit radius R has simultaneously infinite and zero magnitudes, or is resolved into different radii having infinite and zero magnitudes:

$$(R = \infty)_E(R = 0)_{E'} \qquad \text{or} \qquad R = \infty \times 0 = 1 \qquad \text{or} \qquad R = \infty / 0 = 1.$$

In opposition to the hermetic infinite sphere, whose simple unit radius admits at one time uniquely an infinite magnitude that reduces the infinite sphere into a uniquely infinite Euclidean plane with no limiting boundary or with an imaginary limiting boundary, the radius of the Zenonian infinite sphere is a complex whole that under the action of contrary forces, is resolved into infinite and zero radii. Mutually neutralized, the opposite radii give rise to a balanced and permanent real universe of unit radius R, free of all forces by virtue of comprising the totality of forces. Thus, under the simultaneous actions of contrary shrinking and stretching forces, the real physical universe is infinitely extended without disappearing into an infinitely diluted line and infinitely contracted without disappearing into an infinitely compressed point.

With the real limiting point b enveloping infinity and at the same time existing everywhere in infinity, the evolutionary and hierarchical model of the universe and its cosmic singularity collapses. In fact, the hierarchical model of the universe exists insofar we perceive with our finite individual senses the different parts of the world, say the inner infinite Euclidean plane a_n and the outer limiting point b of the universe, as contradictory parts existing successively along the Euclidean line of sight. Thus, according to the big bang evolutionary theory, at one and unique moment in the past called time zero the universe was squeezed into an extensionless and dimensionless limiting point b, or point-fire. At any moment later and under the action of the stretching force of repulsive gravity, an infinitely extended and infinite-dimensional Euclidean universe a_n has been generated from this limiting point b in conformity with the deductive temporal order $b < a_n$, where the limiting point b occurs before the Euclidean universe a_n and the Euclidean universe a_n occurs after the limiting point b.

If we replicate successively and without limit the above temporal order, we obtain the following infinite sequence of cycles of expansion and contraction of the time-conditioned universe under the conflicting actions of stretching and shrinking forces, which is an indefinite recurrence of deductive and inductive temporal orders:

$$b < a_n < b < a_n < b < a_n . . .$$

The deductive temporal order $b < a_n$ states that the transcendent limiting point b—the point-fire (popularly called the big bang) occurring before and beyond the time-conditioned Euclidean universe a_n—was the absolute origin or past of the expanding Euclidean universe a_n and that the expanding Euclidean universe a_n was generated from the limiting point b. On the other hand, the converse inductive temporal (inductive) order $a_n < b$ states that the transcendent limiting point b—the point-fire (popularly called the big crunch) occurring after and beyond the time-conditioned Euclidean universe a_n—will be the absolute end or future of the contracting Euclidean universe a_n, and that the limiting point b will be generated from the contracting Euclidean universe a_n. The time interval between the two conflagrations (point-fires) constitutes the finite age of the time-conditioned Euclidean universe a_n, which the ancient Greeks called the Big Year.[7]

As we have shown in chapters 1 and 2, insofar as there is a discontinuous inequality and temporal order between the parts of the physical universe—say, between the Euclidean universe a_n and its limiting point b—any continuous motion between them, such as the generation of a_n from b and of b from a_n, is impossible. It follows that the above hierarchical model of the universe characterized by the discontinuous succession of its contradictory parts a_n and b, which generates the Euclidean sequence of cycles of expansion and contraction, of deductive and inductive temporal orders, can in no way be the true model of the physical universe receptive of continuous motion between its parts.

What is left, then, as the true model of the real physical universe is the non-hierarchical model where the limiting point b and the Euclidean universe a_n are continuous equal parts of the real infinite sphere. In this case, the limiting point b loses its chronological and transcendent aspect, as it is neither before nor after the Euclidean universe a_n, but instead coexists with and is equivalent to the Euclidean universe in conformity with the synthetic equivalence $a_n = b$. What remains, therefore, is uniquely the spatial and transcendental aspect of the limiting point b, where b is both immanent to and beyond the Euclidean universe a_n of sensible parts. In this sense, our intellectual faculty of infinite synthetic reason conceives the limiting point b not as the chronological origin (or end) of the universe but rather as the geometric principle whose infinite curvature unifies all oppositions of the inner Euclidean part of the real physical universe that our finite individual senses perceive as impossible to unify. The logical manifestation of this geometric principle that governs the limiting boundary of the infinite sphere is the synthetic equivalence principle, which stipulates the equality of all opposites on the infinite sphere.

Thus, opposites that are impossible to reconcile in the inner Euclidean part of the infinite sphere are necessarily reconciled on the infinite sphere. This reconciliation is accomplished by the infinite sphere's limiting point b considered as the unifying principle of all oppositions of the inner Euclidean world. For example, the irreconcilable opposition *flat* and *curved* is reconciled on the complex limiting boundary of the infinite sphere defined as the product of infinitely great and infinitely small spheres. Accordingly, anything located on the complex limiting boundary of the rotating infinite sphere moves along two paths at the same time: i) along the rectilinear path where under the stretching force of repulsive gravity it accelerates away from the starting point and center a in a boundless Euclidean plane, and ii) along the curved path where under the shrinking force of attractive gravity it moves around the starting point and center a with a maximum speed in a bounded sphere (Figure 3.4).

Ultimately, anything that participates in the rotation of the infinite sphere is both away from the starting point a and at the starting point a, thereby showing that the rotating infinite sphere is both moving and at rest, boundless and bounded, open and closed, without absurdity. However, if moving around the circle is to be "both never and always at a starting point. . . (that is why a sphere is both moving and at rest. . .)," we conclude that every rotating

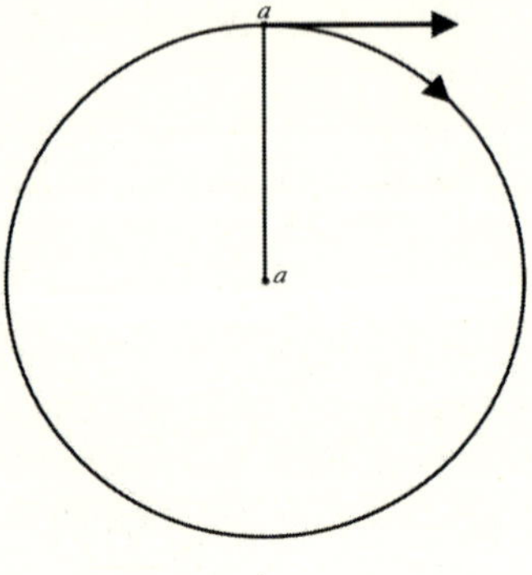

Figure 3.4. Any physical body *a* moving around a circle of positive curvature is moving both i) in a straight line away from the starting point *a* in a boundless universe and ii) in a curved line around the starting point *a*, which is equivalent to being at rest at the starting point *a* in a bounded universe.

finite sphere is essentially in its deepest reality a rotating *infinite* sphere.[8] This means that the immanent nature of circular motion or, more comprehensively, of the sphere in circular motion is not simple, as Aristotle and the entire Aristotelian physics assumed, but instead complex involving the synthetic unity of opposites—of flat and curved, of open and closed, of motion and rest.

After a journey of 15 billion light-years, in the metaphysics of infinity we have finally reached the limiting boundary of the universe where we can answer Archytas's question about whether or not it would be possible at the end of the world to stretch out our hand. The answer is complex because of the complex nature of the limiting boundary of the physical universe. We can stretch out our hand horizontally along the Euclidean line of sight *E*, which is the straight line where we move at an accelerating speed and empirically confirm the assumption that our running universe is open and boundless with respect to space and that it will never halt. However, if we stretch out our hand at a right angle to the Euclidean line of sight *E*, which is the vertical non-Euclidean line of sight *E* 'where we move at a maximum speed, then our hand is infinitely compressed or curved. We are led, then, to empirically confirm the assumption that the running universe is closed and bounded with respect to extent and that it will halt.

Our future behavior as a simple individual located inside the Euclidean part of the physical universe is intrinsically unpredictable. In fact, we are incapable of predicting whether it is possible at the end of the universe to stretch out our hand. However, if we sit on the limiting boundary of the universe where the universe is self-divided into two coexisting flat and infinitely curved parts, we have the power to stretch out and not to stretch out our hand alternately if our body is simple and indivisible or simultaneously if our body is complex and divisible into many bodies. We can then predict with certainty the totality of the alternatives, because we realize all of them. Predicting with certainty the totality of alternatives generates a positive indetermination, which is not a sign of ignorance; on the contrary, it is a sign of our complete knowledge of the complex physical universe, which is simultaneously open and closed, in motion and at rest, divided and undivided. We have negative indetermination when the impossibility of determining (predicting) one of the alternatives expresses our ignorance of the physical universe, which, assumed to be a simple individual, is either open or closed, either in motion or at rest, either divided or undivided.

NOTES

1. In his *Physics* III (4), 203 a_{14} – $a_{16,}$ Aristotle mentions that "Plato has two infinites, the great and the small." Thus, Plato conceives the infinite as a complex whole, which he divides into symmetric opposite infinites: namely, the great infinite according to extension and its inverse, the small infinite according to division. Modern mathematicians designate the first great infinite, which is the absolute maximum according to extension, by ∞ and the second inverse infinite, which is the absolute maximum according to division, called the absolute minimum or zero, by $1/\infty = 0$. Metaphysicians sometimes call the great infinite *extensive infinite* and the small infinite *intensive infinite* or *infinite of position.*

2. *On the Heavens,* trans. J.Tricot (Paris: Librairie Philosophique J. Vrin, 1996), I, 5, 271*b*, 25–30. Indeed, in substance, Aristotle's argument runs as follows: If the rotating body is infinite, then the distance between its radii is infinite. Because it is impossible to traverse the infinite, it follows that it is impossible for a body with an infinite surface to accomplish an infinite circular movement. Aristotle's conclusion is true, insofar as the infinite distance between the radii, and therefore the infinite surface of the body, is simple and indivisible. However, if we choose the Platonic way and divide the one infinite distance between the radii into two infinite distances—the inner infinite distance and the outer zero distance—then combined and mutually neutralized they enable the infinite body to accomplish a circular movement in a finite time, say in one instant, despite its infinite size. If we then assume that the infinite surface of the body is complex and divisible into contrary parts, into infinite and zero surfaces, then Aristotle's conclusion is false.

3. See G.S. Kirk, J.E. Raven, and M. Schofield, *The Presocratic Philosophers*, 267–269.

4. According to Aristotle, the activity of immobility is the immanent and absolute activity of the First Immobile Mover, who imparts maximum motion to the world without being moved. See his *Ethica Nicomachea* , VII, trans. W. Ross (London: Oxford University Press, 1975), 14, 1154^b 27.

5. For example, we can construct the infinite sphere having both infinite and zero dimensions as follows: We start with *a* taken as the center of the 0-D sphere of radius a_0 embedded in 1-D Euclidean space. Subsequently, we conceive this 1-D Euclidean space as a tiny part of a 1-D sphere of center *a* and radius a_1 embedded in 2-D Euclid-

ean space. Subsequently, we conceive this 2-D Euclidean space as a tiny part of a 2-D sphere of center *a* and radius a_2 embedded in a 3-D Euclidean space. Subsequently, we conceive this 3-D Euclidean space as a tiny part of a 3-D sphere of center *a* and radius a_3 embedded in 4-D Euclidean space. Replicating this process an infinite number of times, we produce an infinite sequence of homocentric spheres of increasing radii a_n and number of dimensions *n*. In fact, we can assert that the radius and number of dimensions of the expanding spheres increase proportionally to their distance from their common starting point *a*. Ultimately, at the limit *b*, we reach the infinite sphere resolved into two coexisting spheres: i) the inner ∞-D sphere of center *a* and radius a_∞ of infinite magnitude, which is equivalent to the ∞-D Euclidean plane and ii) the outer 0-D sphere of center *a* and radius a_0 of zero magnitude, which is equivalent to the limiting point *b* circumscribing the ∞-D Euclidean plane. The 0-D sphere is therefore both the beginning and the end of all magnitudes, the contained part embedded in 1-D Euclidean plane and the containing whole enveloping the ∞-D Euclidean plane.

6. If the radius of the infinitely great sphere is infinite $R = \infty$, then the curvature of its limiting boundary is $K = 1/R^2 = 1/\infty = 0$ and the limiting boundary is flat, or equal to the Euclidean plane of zero curvature and force. If we regard zero curvature as negative or imaginary infinite curvature such as $K = 0 = \infty'$, then the flat limiting boundary has a negative or imaginary infinite curvature, which is the curvature of the hyperbolic point and constitutes the imaginary limiting boundary of the Euclidean plane. The hyperbolic point is the inner concave part of the infinite sphere, which curves inward and away from the Euclidean eye. Because the absence of curvature and force can be presented as a negative or imaginary curvature and force, we present the straight line of zero curvature and force as a hyperbolic or concave point of negative infinite curvature and force.

7. The idea of conflagration (εκπύρωσις), the periodical generation and destruction of the universe by fire, although a common idea circulating among the ancient thinkers of Persia, Egypt, India and Ionian Greece, was vividly defended by the Ionian Greek philosopher Heraclitus (sixth century BCE). According to Heraclitus, the universe is born from fire and destroyed by fire alternately and indefinitely in time under the conflicting actions of discord and concord, or as the cosmologists of today would say, under the contradictory stretching and shrinking forces of repulsive gravity and attractive gravity corresponding, respectively, to the conflicting actions of discord and concord.

8. Aristotle, *Physics*, trans. H. Carteron (Paris: Les Belles Lettres, 1986), VIII (9), $265b_2$.

Chapter Four

What Is Quantity?

Even as the finite encloses an infinite series
And in the unlimited limits appear
So the soul of immensity dwells in minutia
And in the narrowest limits, no limits inhere.

—Jacob Bernoulli

The immanent complexity and indeterminacy of the unit distance *ab* possessing at the same time contrary magnitudes, that is, finite and infinite magnitudes, leads us to rethink the Euclidean-Aristotelian concept of quantity, according to which no quantity admits contraries.[1] It is this precisely analytic property of quantity that leads us to define *quantity* as anything that has at one time a unique magnitude and obeys analytic principles of organization. We call this univocal or simple quantity an *individual*. The geometry of the individual quantity is perceived by the observer's individual senses at rest on earth as being Euclidean free of force.

In fact, only in Euclidean space, where there is approximately no force to disturb (expand/contract) the measuring apparatus and hence the given quantity to be measured, can the observer establish exactly at one moment the unique magnitude of the given quantity. The Euclidean-Aristotelian concept of quantity relies, therefore, on two fundamental empirical assumptions: i) that any quantity of a given kind is a simple indivisible thing—an individual—admitting at one moment a unique magnitude, and ii) that the geometry of this simple quantity is Euclidean of zero force.

Through our study of the problem of motion and its correlative solution, however, we have discovered that the unit distance *ab* is neither a simple quantity nor a Euclidean one. In fact, we have discovered that continuous motion from *a* to *b* separated by an actually infinite number of parts of positive magnitude is impossible, if we assume that the unit distance *ab* is a simple quantity obeying analytic principles of being and a Euclidean quantity having at one time a unique magnitude, which is either finite or infinite. More specifically, we have discovered that in order to effect the continuous passage from *a* to *b*, the unit distance *ab* must admit at the same time contrary magnitudes—namely, finite and infinite magnitudes (according to extension and division). In other words, the unit distance *ab* must be in its essence (τό κατ'αὐτό) and deepest physical reality a non-Euclidean complex and indeterminate quantity that is both finite and infinite and therefore has the totality of magnitudes, and yet it is neither finite nor infinite and has no magnitude at all. We call the *universe* or *maximum* or *absolute* quantity this Zenonian unit distance *ab* in which all opposite determinations coincide, and which we identify with the infinite sphere's non-Euclidean radius *ab,* which is the arc *ab* of its great circle.

When, however, we look at this real physical quantity—namely, at the real unit distance *ab* conceived by our synthetic reason as a universe of magnitudes—we perceive it as something else, that is, as an individual sensible unit distance *ab* having at one time a unique magnitude that we identify with the infinite sphere's inner Euclidean radius *ab*. We say, then, that our individual senses suffer from *perceptual illusion* when we perceive something as something else, the universe *as if* it is an individual, the real quantity *as if* it is a sensible quantity, and that this perceptual illusion is the result of the imperceptible contraction or constraint exercised by our Euclidean individual senses on the real quantity. In fact, this imperceptible constraint causes the collapse of the universe of magnitudes belonging to the real unit distance *ab* into a unique magnitude belonging to the individual sensible unit distance *ab*. Before observation, the unit distance *ab* was a universe admitting contrary magnitudes, that is, both finite and infinite magnitudes, and in this sense it was a real physical quantity existing in itself (τό κατ' αὐτό ποσόν) and independently of the individual observer. After observation, however, the same unit distance *ab* is contracted by our individual senses into an individual sensible unit distance *ab* admitting at one time a unique magnitude that is either finite or infinite. Because it lacks the totality of magnitudes, it is an incomplete sensible quantity—a modified image or appearance of the original real quantity.

The choice of observing a unique magnitude among alternative magnitudes is completely arbitrary and depends on the circumstances of our individual perception: for example, on our position, motion, and biological structure such as the size of our brain cell or the frequency of its firing. This choice, which quantum physicists call *de-coherence* because it disrupts the unity and coherence of the universe of magnitudes, leads us to assign to the sensible unit distance *ab* the extrinsic quality of being accidental—that is, of being neither necessary nor constant and hence a variable that has at one time a unique magnitude and at different times different magnitudes. We thus assimilate the Euclidean sensible unit distance *ab* with the Euclidean variable a_n located inside the infinite sphere.

On the other hand, we assign to the real unit distance *ab,* which has at one time contrary magnitudes or a universe of magnitudes, the inherent quality of being a substance: that is, of being necessary, constant and independent of the constraining effects of individual perception. We subsequently number this real unit distance *ab*, which we defined as the product or ratio of infinite and zero magnitudes and identified with the infinite sphere's non-Euclidean unit radius (arc) *ab*, by the real infinite whole 1. We then consider this real numeric quantity 1 as the *first quantity,* which is chronologically and ontologically prior to the individual sensible quantity a_n. And we regard the individual sensible quantity a_n as the *second quantity*, which is chronologically and ontologically posterior to the first quantity. Ontological priority refers to things that are themselves independent but in which the being of others depends. Thus, the real infinite whole 1 is independent of its sensible part a_n, whereas the sensible part a_n depends for its being on the real whole 1.

Based on this distinction, which Plato used, we obtain the following chronological and ontological deductive order: $1 < a_n$, where the real quantity 1 is chronologically before and ontologically prior to the individual sensible quantity a_n having a unique magnitude, and which is the modified (reduced) version of the real quantity 1 having a universe of magnitudes. It follows that the assumption that there is no disturbance, no force that modifies (constrains) the measurement of the real unit distance *ab*, is false. In fact, our very act of individual perception exercises an imperceptible force of contraction over the real physical quantity 1, so what we perceive and measure is not the real quantity 1, but rather the reduced sensible quantity a_n, which is the inexact magnitude of the real quantity 1. It is obvious that insofar as our individual perception of things generates the chronological order $1 < a_n$, where we perceive the real quantity 1 as something else—namely, as a sensible quantity a_n regarded as a modified (contracted) version of the original real quantity 1—our individual perception will be synonymous with perceptual illusion or with the gravitational force of contraction.

We also suffer from *error* when, as individual observers imprisoned in the Platonic cave of our senses, we take falsity as truth, appearances for real things, and the sensible quantity a_n as the real quantity 1. We are also in error when we consider the unique magnitude of the sensible quantity a_n to be the exact magnitude of the real physical quantity 1; the Euclidean geometry of the individual sensible things to be the natural geometry of the real physical things; and Aristotle's analytic principles of thought, which are abstractions of our individual senses, to be the immanent principles of the real physical world and thus the ultimate sources of knowledge. Finally, insofar as the infinitely small difference $1- a_n = e$, which defines the error between the real quantity 1 and the sensible quantity a_n, is in reality an infinitely great difference in which a_n is as far off from 1 as the least finite number a_0 (see chapter 1), the infinitely small error from which we suffer as individual observers who regard the sensible quantity a_n as if it is the real quantity 1 is in reality an infinitely great error.

Now let us assume that the analytic principle of the excluded middle is the ontological principle that governs the real physical universe. According to this principle the real physical universe (or every real physical being) is either a or b:

$$(a + b),$$

where the sign of addition or disjunction (+) designates either/or. It follows that one of the conflicting opposites must be true. We call *analytic ontology* the ontological doctrine affirming that the real physical universe (or every real physical being) is a simple individual admitting at one time a unique determination , namely either determination a or determination b, and therefore verifies analytic principles of being.

If a stands for finite and a'(not-a), which we call b, stands for infinite (not-finite), let us determine the ontological and gnosiological consequences of the above assumption with respect to the alternatively affirmed opposites finite/infinite. We postulate, therefore, the following propositions:

p = the physical universe is a real being existing in itself (τό κατ̓ αὑτό ὄν)
q = the physical universe is either finite or infinite

Grounded in the deductive rule of *modus ponens* $p < q$, we affirm that if the physical universe is a real being existing in itself, then it is either finite or infinite. But we have discovered (see chapter 3) through our faculty of synthetic universal reason that the physical universe's unit radius *ab* has both finite and infinite magnitudes (according to extension and division). When they are mutually negated, they assign to the unit radius *ab* neither a finite magnitude nor an infinite magnitude and hence no magnitude at all. This leads us to conclude, according to the deductive rule of *modus tollens* $q' < p'$, that if it is false that the physical universe is either finite or infinite, then it is false that the physical universe is a real being existing in itself. It follows, then, that the physical universe is an unreal, apparent, or impossible being.

We note here that the impossibility of determining one of the two alternatives—that is, deciding whether the physical universe is finite or infinite with respect to magnitude—indicates on the ontological level the unreality, impossibility, or appearance of the physical universe. On the other hand, assuming that the physical universe is a real being, the impossibility of deciding one of the two alternatives equally indicates on the epistemological level our absolute ignorance of the real being of the physical universe, which is assumed by our analytic understanding in association with our individual senses to be simple and determinate and therefore either finite or infinite with respect to magnitude.

It is clear that the confrontation between the assumption that the analytic principle of the excluded third is the immanent principle of the real being of the physical universe and the conclusion of our synthetic universal reason denying the above assumption leads us directly to the doctrines of ontological and epistemological nihilism. By ontological nihilism we mean the ontological doctrine of *irrealism,* which affirms the unreality of the physical universe, whereas by epistemological nihilism we mean the epistemological doctrine of *skepticism*, which affirms, given the real being of the physical universe, our irrefutable ignorance of its simple and determinate nature.

The impossibility of knowing the real nature of the physical universe, which our faculty of finite analytic understanding improperly assumed to be simple and determinate obeying analytic principles, gave rise to different doctrines of skepticism during the growth of science. For example, it engendered Nicholas of Cusa's *De Docta Ignorantia* (*On Learned Ignorance*), which affirms the impossibility of formulating a simple, univocal representation of the physical universe; Kant's transcendental idealism, which affirms the impossibility of knowing the real physical universe as a simple whole existing in itself, which is either finite or infinite;[2] Turing's unpredictability problem, which affirms the impossibility of determining whether a given program in the machine will terminate or continue running for ever; and Gödel's incompleteness theorem of truth, which affirms the impossibility of stating a simple proposition about the physical universe that is either true or false, either proved or disproved. All these different aspects of skepticism are derived from one general impossibility—the impossibility of our infinite reason to give a unique determinate answer that is either answer *a* or answer *b* with respect to this simple ontological question: "Is the real physical universe (or any real physical being) *a* or *b*?"

Because the impossibility or negation of the determinate answer implies the negation of the analytic principle of the excluded middle upon which the determinate question and answer are based, the physical universe and our synthetic reason are necessarily led to formulate an indeterminate answer that is neither answer *a* nor answer *b*. This indeterminate answer, however, is negatively interpreted by our analytic understanding, which, following the skeptical line of reasoning, claims that it reveals either the unreality of the physical universe or, given its reality, our absolute ignorance of its simple determinate nature (Kant).

We can escape this impasse of ontological and epistemological nihilism—which is the consequence of the confrontation between an analytic ontology of the being and our synthetic universal reason, considered as the faculty of thinking of the real physical universe as a complex infinite whole—if we replace this analytic ontology by its negation. We call *synthetic ontology* the ontological doctrine affirming that any real physical being existing in itself is a complex indeterminate universe that, has at one time equal and opposite determinations called contraries and verifies synthetic principles of being. Thus, with respect to magnitude, the real physical universe is a complex being that is both finite and infinite (according to extension and division); it is equally an indeterminate being that, by virtue of negating the extremes, is neither finite nor infinite.

In line with these considerations, let us see what happens if we cease to consider the analytic principle of the excluded middle as the immanent principle of the real physical universe and consider in its place its negation: the synthetic principle of the included middle. According to this principle, the real physical universe (or every real physical being) is neither (not-either) *a* nor *b*:

$$(a + b)',$$

where (+) designates either/or and (′) designates negation (not), falsity, impossibility. It follows that both extreme opposites must be false: $(a + b)' = a'b'$. By negating both extremes we obtain an impartial, and indeterminate physical universe that operates as a temperate middle point equidistant from the determination of extremes and hence free of determination.

Let us see the ontological and gnosiological consequences of the above assumption with respect to the simultaneously negated extreme opposites finite/infinite. We stipulate the following propositions:

p = the physical universe is a real being existing in itself
q = the physical universe is neither finite nor infinite

Grounded in the deductive rule of *modus ponens* $p < q$, we affirm that if the physical universe is a real being existing in itself, then the physical universe is neither finite nor infinite. Indeed, by means of our faculty of synthetic universal reason (see chapter 3), we discovered that the physical universe's unit radius *ab* has both finite and infinite magnitudes (according to extension and division), which, when they are mutually negated, they assign to the unit radius *ab* neither a finite magnitude nor an infinite magnitude and hence no magnitude at all. This leads us to conclude that, according to the inductive rule of conversion $q < p$, if it is true that the physical universe is neither finite nor infinite, then it is true that the physical universe is a real being existing in itself.

We must note that the impossibility of deciding one of the two alternatives, for example either the finite magnitude or the infinite magnitude, shows on the ontological level the inherently complex and indeterminate nature of the physical universe and on the gnosiological level our complete and maximum knowledge of its complex and indeterminate nature. Indeed, according to the synthetic principle of the included middle, the physical universe has neither a finite nor an infinite magnitude and thus is an indeterminate being having no magnitude at all. And yet according to the finite-infinite equivalence principle, the physical universe has both finite and infinite magnitudes and thus is a complex being having a universe of magnitudes constituting the absolute and real quantity 1 that grounds any indefinitely varying sensible quantity a_n of the Euclidean world of our individual sense. We call ontological *realism* the doctrine affirming that the physical universe is a real being existing in itself and independently of the individual observer; we call epistemological *gnosticism* the doctrine affirming the necessity of attaining complete and absolute knowledge of the complex and indeterminate nature of the real physical universe.

By changing the analytic ontology of the real physical universe into a synthetic ontology in which the synthetic principle of the included middle governs the real physical universe, we transform the impasse of ontological and epistemological nihilism into the true answer given by the doctrines of ontological realism and epistemological gnosticism. Thus, the reply to the simple question "Is the real physical universe *a* or *b*?" is not a simple answer in which the selection of one of the alternatives begins a viciously endless argument that generates an unsolvable conflict. If the answer is the logical negation of the question, then it is necessary that the answer be complex and indeterminate, that is, outside the simple structure of the question. It follows that the complete and genuine answer to the above simple question is that the real physical universe (or every real physical being) is both *a* and *b* and neither (not-either) *a* nor *b*:

$$(a = b)\,(a + b)', \qquad \text{or} \qquad (a \times b)\,(a + b)',$$

where the sign of equality, unity or simultaneity (=) and the sign of multiplication or conjunction (×) stand for both, and the sign of addition or disjunction (+) stands for either/or.

Indeed, according to the synthetic principle of the unity, equality, or conjunction of extreme opposites, every *a* is simultaneously a *b* and every *b* is simultaneously an *a*. In other words, the physical universe is both *a* and *b*. The particular form of the synthetic principle of the unity of opposites is the finite-infinite equivalence principle, which stipulates that everything finite is infinite and everything infinite is finite, and that the real physical universe is both finite and infinite. According to the synthetic principle of the included middle, the real physical universe is an indeterminate whole that is neither (not either) *a* nor *b*, neither finite nor infinite. Because there is no sufficient reason to choose answer *a* over answer *b*, or answer *b* over answer *a,* the real physical universe is equidistant from the extremes and hence identical to the temperate middle point, the neutral and free point, ensuring the unity and balance of the extreme opposites and, therefore, the justice and permanence of the physical universe.

The logical and ontological distinction between the *individual* and the *universe* is traced back to Aristotle and has become throughout the history of philosophy a classical distinction.[3] For Aristotle, the individual satisfies analytic principles of thought, which are in turn abstractions of our individual senses. For example, according to the analytic principle of contradiction, nothing is both *a* and *b*, which means that if a thing is both *a* and *b*, then it is an impossible thing —a nothing. Opposite parts or determinations that cannot coexist in the same being experienced as an individual are called *contradictories.* On the other hand, opposite parts or determinations that coexist in the same being conceived as a universe, whole or maximum are called *contraries*, and the maximum difference and distance between contraries we call *contrariety.*

Because contradictory determinations are conflicting opposites that cannot coexist in the same individual, they exist successively in the same individual in conformity with the analytic principle of temporal order. Thus, at one moment the individual possesses a unique determination or contradictory determinations and at different moments consecutive determinations. If at one moment the individual is *a,* then at the consecutive moment the individual is *b*. It follows that the individual is anything simple, determinate, and time-conditioned that can neither be extended nor divided into coexisting opposite determinations. On the other hand, the universe, whole, or maximum is anything complex, indeterminate, and timeless that can be extended by being divided into opposite parts *a* and *b* at the same time. When these coexisting opposites define the diameter of the universe, they are diametrically opposite and inseparable poles *a* and *b* of a universe regarded as a dipole magnet. When they define the radius of the universe, they are spatial proto-contraries, such as the center *a* and the limiting boundary *b*, that determine the maximum distance between points within the universe and from which we can derive, according to Aristotle, all other contraries.[4]

Aristotle called *syllable* (συλλαβή) the unity of different elements or elementary parts *a* and *b*.[5] The word syllable is derived from the verb συν + λαμβάνω, which means to take together. We take together the different parts *a* and *b* in order to compose a complex unit whole or universe in which its different parts are united and continuous. The unity of opposites or the unity of multiplicity can be expressed i) by the sign of multiplication or conjunction (×) denoting *both*, and ii) by the sign of equality (=) denoting *if and only if*, identity, simultaneity, or the connecting word *is*. For example the equality:

$$a = b$$

affirms the simultaneous and interdependent existence of opposite parts, that *a* exists if, and only if, *b* exists and vice versa regardless of their difference and distance, whereas the conjunction:

$$a \times b = 1$$

affirms that every *a* has an opposite $b = a'$ (*b* is not-*a*) such that *a* and *b* form an independent complex unitary whole—in which they are contraries, symmetric or polar opposites, reciprocals, correlatives, duals—that cannot be separated. Thus, we can superpose or coincide any opposite part with the other opposite part by means of a 180-degree rotation about the whole's center. Because each opposite part is the negation that completes (and not contradicts or destroys) the other opposite part, for example $a = b'$ and $b = a'$, we can also call each opposite part a *complement* of the other part. If the conjunction sign (×) denotes the intersection of the different parts *a* and *b*, then their logical product ensures their penetrability: that is, two different parts occupying the same place at the same time.[6] The same place and time occupied by the two different parts *a* and *b* constitute that which is common and universal between different parts.

The complex universe or syllable *ab* has, as a living spherical whole, the power to replicate itself by division so the self-divided universe is made of simultaneous universes. Thus, when the original universe or whole *ab* divides itself into two parts, we obtain two new wholes, each replicating the original coexisting parts *a* and *b*. When each of these two wholes is divided into two parts we obtain four wholes, each replicating the original coexisting parts *a* and *b*. We note here that the infinite division of the whole into parts, which in turn are wholes themselves, shows that living wholes are resistant to damage and in this sense are indestructible and timeless. In spite of the division of the whole into parts, the whole remains indivisible and is composed of indivisible wholes, which are simultaneously divided, and not of indivisible parts, which are not divided. Now, by dividing an infinite number of times the original spherical living whole, we infinitely double the original whole such that ultimately we obtain an infinitely divided and doubled whole $1/2^{\infty} \times 2^{\infty}$ equal to the real whole $1 = 0 \times \infty$, which we consider to be coextensive to the real physical universe, defined as the whole of wholes or the universe of universes (Figure 4.1).

Let the geometric figure of the real physical universe be the real infinite sphere of center *a* and maximum radius *R* equal to the constant real infinite whole $1 = \infty \times 0$. However, when we look at the physical universe, defined as the total sum of parts, we see it differently, that is, as the *observable* or *sensible universe*, defined as the partial sum of parts, whose geometry is the imaginary infinite sphere of center *a* and radius equal to the sensible part a_n varying without limit. We distinguish, therefore, two kinds of quantities: i) the Euclidean sensible quantity a_n, which is the sensible universe or any sensible quantity of a given kind that belongs to the sensible world, and ii) the non-Euclidean real quantity $1 = \infty \times 0$, which is the real physical universe and constitutes the maximum physical reality of any sensible quantity a_n belonging to the sensible universe. In turn, we consider this real quantity 1—which is the *first origin* and *principle* of the sensible quantity a_n according to the chronological *deductive order* $1 < a_n$—as the *final end* of the sensible quantity a_n according to the following chronological *inductive order*:

$$a_n = (ab)'\,(a + b) < 1 = (ab)(a + b)'.$$

This inductive order states that the sensible universe or sensible quantity a_n, which is before, less than, and ontologically inferior to the real quantity 1, verifies analytic principles of existence, whereas the real physical universe or real quantity 1, which is after, greater than, and ontologically superior to the sensible quantity a_n, verifies synthetic principles of being. With respect to the sensible quantity a_n, for example, the analytic principle of contradiction states that no sensible quantity is both *a* and *b*, both finite and infinite (according to extension and division), whereas the analytic principle of the excluded middle states that any sensible quantity is either *a* or *b*, either finite or infinite. The sensible quantity a_n relies on two fundamental Aristotelian assumptions: i) that any sensible quantity a_n of a given kind is a simple individual admitting at one time a unique magnitude that renders it measurable

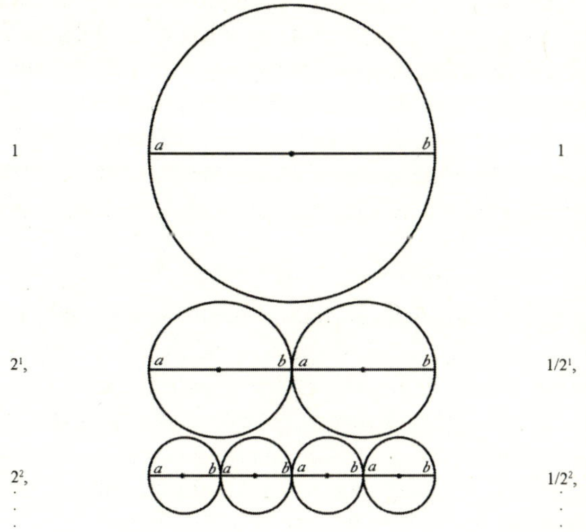

Figure 4.1. The living universe or whole represented by the sphere is a dipole magnet having two equal and opposite poles *a* and *b*. When we divide the original spherical living whole into two parts, we obtain two new wholes, each replicating the original coexisting poles *a* and *b*. Continuing in this way, we obtain a complex, infinite sequence of divided and doubled wholes $1/2^n \times 2^n$, which has as its maximum limit the infinitely divided and doubled real whole $1 = 1/2^\infty \times 2^\infty = 0 \times \infty$, defined as the whole of wholes or the universe of universes.

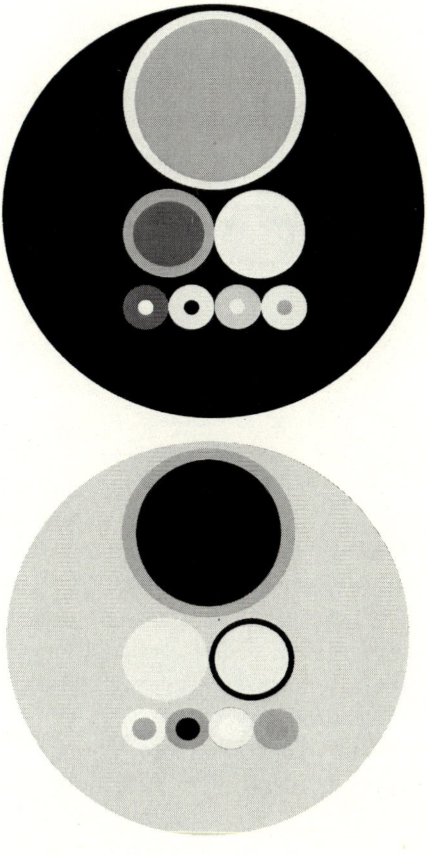

Figure 4.2. Artistic expression of the generation of multiplicity within the world by dividing the spherical whole into parts, which are themselves spherical wholes. Realization: Setareh Korkchi.

or computable, and ii) that the geometry of this simple finite quantity located in the inner part of the infinite sphere is Euclidean. However, as soon as the progressing sensible quantity a_n attains its maximum limit *b* and discovers its greatest, highest, and deepest reality—that of being a maximum quantity, which is the real physical quantity 1 located on the limiting boundary *b* of the infinite sphere—it replaces its accidental analytic principles of existence by synthetic principles of being, which are the true and immanent principles of the real physical universe.

Thought of as a universe of magnitudes, the real quantity 1 is, according to the synthetic principle of the unity (or coincidence) of opposites, both *a* and *b*, both finite and infinite (according to extension and division). On the other hand, the real quantity 1 is, according to the synthetic principle of the included middle, neither (not-either) *a* nor *b*, neither finite nor infinite. The coincidence of opposites in the same real quantity 1 is an absurdity if we regard the quantity as an individual belonging to the Euclidean plane a_n, whereas it is a logical necessity if we regard the real quantity 1 as a universe or a maximum belonging to the infinite sphere 1.

The real physical quantity 1 relies on two fundamental non-Aristotelian assumptions: that any sensible quantity a_n of a given kind is in its highest and deepest reality a real physical quantity 1, which is a complex indeterminate universe or a maximum, having at one time contrary magnitudes and ii) that the geometry of this real physical quantity 1 located on the infinite sphere is non-Euclidean free of measure and computation by comprising the totality of measures and computations.

COMPLEX QUANTITIES IN BODIES MOVING AT THE SPEED OF LIGHT

Let us imagine that at one moment in your life, you decided to explore the physical universe and discover the ultimate reality of sensible things on its limiting boundary by following the receding motion of the galaxies relative to earth. Taking control of the cosmic stretching force of expansion (centrifugal force), which causes the receding motion of the galaxies, you yourself, or, more precisely, your vibrating brain, embark on this eccentric exploratory odyssey. According to the Hubble law $v = kd$, where *k* is a constant of proportionality and *d* is the distance from earth, your finite brain's speed of recession *v* increases proportionally to its distance from earth, regarded as the starting point *a*.

Until your finite brain has reached the speed of 225.000 kilometers per second, which is 75 percent of the speed of light or 0.75*c*, its mass *m*, and the speed c of its radiated light have their gamma factor $\gamma = 1/\sqrt{1 - v^2/c^2}$ approximately constant and equal to 1. The magnitude of the γ factor tells us how much a quantity of a given kind varies in function of the moving body's speed. Because the cosmic acceleration has no significant effect on the brain's sensible quantities of mass and speed of its radiated light, they remain constant, thereby constituting the brain's *rest mass m* and the light's *rest speed c* (the speed that light has when it moves through constant empty space) within a Euclidean rest frame. In such a quasi-constant Euclidean world, any sensible quantity of a given kind belonging to your finite brain and its radiated light has a simple and determinate meaning. This means that in conformity with the analytic principles that govern our Euclidean observable world, your brain's mass m or the speed c of its radiated light or simply their γ factor admit at one time a constant and unique magnitude that we take as unity. It is therefore a simple indivisible unity.

However, as soon as you reach and surpass the Euclidean limit of 0.875*c*, the cosmic force of expansion is divided into the cosmic forces of expansion and contraction, which cause the symmetric variation of the γ factor according to increase relative to the horizontal axis of expansion and according to decrease relative to the vertical axis of division or contraction.[7] At 0.875*c*, the γ factor is roughly doubled and halved; at 0,968*c*, the γ factor is roughly twice doubled and halved; at 0.992*c*, the γ factor is roughly thrice doubled and halved and so on. Finally, at the maximum speed c, the γ factor is a real infinite whole 1, which is both infinitely doubled and halved with respect to these opposite axes.

We say, then, that at the maximum speed of light, the simple and finite γ factor and any quantity depending upon the γ factor are maximized or infinitized.[8] In fact, the γ factor acquires with respect to the orthogonal axes infinite and zero magnitudes, which transform it into a complex indeterminate quantity equal to the real $1 = \infty \times 0$ (Figure 4.3).

It follows that when your receding finite brain accelerates at the maximum speed of light at the limiting boundary b of the infinite universe, its mass m and the speed c of its radiated light are maximally and symmetrically varied, acquiring infinite and zero magnitudes at the same time. We say, then, that your radiating brain's *sensible quantities m* and c having at one time a unique magnitude, become at the limiting boundary b of the infinite universe *real quantities m′* and c', having at the same time contrary magnitudes, namely infinite and zero magnitudes—the product of which is unity, a complex indeterminate unity. We have therefore

$$m' = m \times \gamma \times 1/\gamma = 1 \qquad \text{and} \qquad c' = c \times 1/\gamma \times \gamma = 1.$$

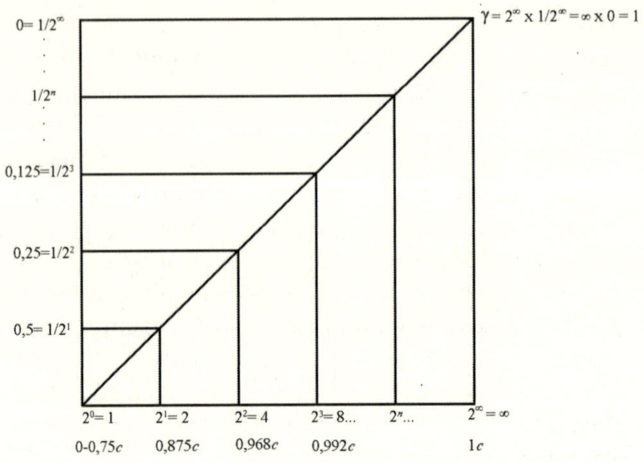

Figure 4.3. The double (two-way) variation of the γ factor caused by the contrary forces of expansion and contraction starts at 0.875 c and takes place along the contrary horizontal and vertical axes. Every expansion by doubling the γ factor along the horizontal axis implies its corresponding contraction by halving it along the vertical axis, and vice versa. Ultimately, when the moving quantity has reached the speed of light c, its γ factor is both infinitely doubled and infinitely halved relative to the horizontal and vertical axes. Because variation is inherently symmetric, it takes place on the diagonal axis, which is the composition of the opposite horizontal and vertical axes, and is self-neutralized so that ultimately the γ factor is equal to the constant real quantity 1, which is defined as the product of infinite and zero magnitudes.

When the $\gamma = 1/\sqrt{1-v^2/c^2}$ factor is maximized or infinitized, that is, becomes equal to infinity at the limiting boundary b of the physical infinite universe, then your brain's real quantities m' and c' have infinite and zero magnitudes at the same time. By substituting, in the formulas of page 42, γ for ∞, we obtain

$$m' = m \times \infty \times 0 = 1 \qquad \text{and} \qquad c' = c \times 0 \times \infty = 1.$$ [9]

Accordingly, if a finite observer on earth were to observe this maximally accelerated brain with a perfect telescope radially, that is, along the horizontal Euclidean line of sight E, that observer would experience your brain as having an infinite mass, an infinite momentum, and, for any given volume, an infinite density. This means that the finite observer on earth would experience your maximally distant brain moving at the speed of light as being maximally compressed to a singularity—that is, to a nothing of *infinite heaviness*. Moreover, the terrestrial observer would experience the speed of your brain's radiated light as being zero and your brain as being a black hole radiating no light, or, to put it another way, radiating light of zero frequency.

The inverse would happen with a finite observer on earth looking at a right angle to the Euclidean line of sight E, that is, along the vertical non-Euclidean line of sight E'. The finite observer would see your brain as having a zero mass and, for any given volume, a zero density. It follows that the terrestrial observer would experience your maximally distant brain moving at the speed of light as being a maximally diluted universe, or a massless photon—that is, a nothing of *infinite lightness*. With respect to the speed of your brain's radiated light, the finite observer would think this speed as being infinite and your brain as being an infinitely luminous star radiating light at infinite frequency.

Although your maximally accelerated brain has broken the light barrier and is, therefore, infinitely compressed with your surrounding space into a singularity, which we assimilate with the universe's outer limiting point b, you continue to exist as usual regardless of your infinite density. This is because your living brain's real density is not infinite density, but rather the complex product of mutually neutralized infinite and zero densities from which the real density 1 emerges. Expressed as a dimensioned quantity, the real density 1 is the density of the water (10^0gram per cubic centimeter) that characterizes any temperate living being located at the middle of the scale of densities, where it is regarded as the intermediate balancing factor between extreme densities.

Thus, looking at your maximally distant brain along the Euclidean line of sight the finite observer on earth would experience your brain as an infinitely compressed singularity. At the same time, the spectrograph of the perfect telescope would detect your brain's water molecules at room temperature. These observations must not be interpreted as contradictory, that is, revealing a conflict between the water density of the living brain and its infinite density due to its speed of light, but instead as complementary, revealing the very complexity of the brain's real quantities.[10]

At what distance from the earth does your accelerating brain reach the limiting boundary b of the infinite universe where it attains the maximum speed of light c and therefore has its finite sensible quantities—for example, its finite unit mass—maximized or infinitized? To put it in a more general way, at what distance from the earth does the limiting point b emerge to partially close the unlimited sequence a_n of our accelerating Euclidean sensible world? At what level of extension of matter or space will we find the greatest body—the infinite universe—moving at the greatest speed?

Let us take the speed-distance equivalence law (Hubble's law) $v = k \times d$, which stipulates that the speed of the moving body is proportional to the distance from its starting point *a* for example, the earth. If the moving body's maximum speed *v* is *c* and the constant of proportionality $k = v/d$ is approximately 3×10^4 meters per second $/1.5 \times 10^{22}$ meters, then the distance $d = v/k$, which corresponds to the moving body's maximum speed $c = 3 \times 10^8$ meters per second, is approximately

$$d = 3 \times 10^8 \div 3 \times 10^4/1.5 \times 10^{22} = 1.5 \times 10^{26} \text{ meters (m)}$$

or 15×10^9 (15 billion) light-years (LY).

We consider this finite distance 1.5×10^{26} meters, which we take as unity, as the maximum distance from the earth that corresponds to your receding brain's speed of light *c* at the limiting boundary *b* of the infinite universe. This speed of light is also the very speed at which the infinite sphere, that is, the body of the infinite universe, rotates about and away from its center and starting point *a*. We in turn identify this finite maximum distance from the earth with the finite maximum unit radius of the infinite sphere, which is divided at the limiting boundary *b* into the inner, infinitely great sphere of infinite radius and the outer, infinitely small sphere of zero radius (see chapter 3). We may then consider the finite length 1.5×10^{26} meters, which is 10^{30} times the middle length 10^{-4} meters (approximately the size of the human egg), to be the finite computable aspect of the universe's infinite radius. At the same time, it's diametrically opposite finite length 1.5×10^{-34} meters, which is 10^{-30} times the middle length 10^{-4} meters, we consider to be the finite computable aspect of the universe's zero radius.

Seen from inside, the infinite universe has an infinite radius, which is the limit of an infinite sequence of self-doubling universes of increasing radii lying on the horizontal axis of the Euclidean line of sight *E*. Seen from the outside, the infinite universe has a zero radius, which is the limit of an infinite sequence of self-halving universes of decreasing radii lying on the vertical axis of the non-Euclidean line of sight E'. Mutually neutralized, these contrary greatest and smallest radii produce the maximum real radius equal to 1 located on the infinite sphere and defined as the product of infinite and zero magnitudes or as the ratio of infinite and zero radii.

Now, these greatest and smallest finite radii are both absolute or constant implying zero variation and relative or variable implying infinite variation. For example, if we multiply the finite magnitude10^{26} by the factor 10^1 an infinite number of times, we obtain the infinite magnitude $10^{26} \times 10^{\infty}$, which although greater than 10^{26}, relative to the inner horizontal axis of the Euclidean line of sight *E*, is equal to 10^{26} relative to the outer vertical axis of the non-Euclidean line of sight E'. We write therefore:

$$(10^{26} \times 10^{\infty} > 10^{26})_E (10^{26} \times 10^{\infty} = 10^{26})_{E'}.$$[11]

This formula states that the infinite magnitude 10^{∞} both increases and does not increase the finite and greatest magnitude 10^{26}.

The same holds true for the smallest magnitude 10^{-34}. If we multiply the finite magnitude 10^{-34} by the factor 10^{-1}an infinite number of times, we obtain an infinite magnitude $10^{-34} \times 10^{-\infty}$, which although less than 10^{-34} relative to the inner horizontal axis of the Euclidean line sight E, is equal to 10^{-34} relative to the outer vertical axis of the non-Euclidean line of sight E'. We write therefore:

$$(10^{-34} \times 10^{-\infty} < 10^{-34})_E (10^{-34} \times 10^{-\infty} = 10^{-34})_{E'}.$$

This formula states that the infinite magnitude $10^{-\infty}$ both decreases and does not decrease the finite and smallest magnitude 10^{-34}.

As we have said, the mathematics of the maximum is synthetic and defies the laws of arithmetic as well as the analytic principles of organization: for example the principles of inequality, comparability, and temporal order. It follows that maximum quantity is not that which cannot be further varied (increased/decreased) and constitutes the analytic definition of the maximum, but rather that which cannot be further varied no matter how much we vary it. Thus, maximum quantity is anything that is both variant and invariant, open and closed, an unlimited line and limiting point with respect to extension and division without contradiction or paradox. This last definition involving the composition of opposites constitutes the synthetic definition of the maximum.

We have defended the thesis that the limiting boundary of the infinite universe rotating at the speed of light is the region where the sensible quantity of a given kind, for example, the finite variable a_n having at one time a unique magnitude, becomes the maximum real quantity—the Zenonian infinite quantity $1= \infty \times 0$, having at one time contrary magnitudes. We may then redefine the maximum real quantity, which is the sensible quantity's highest and deepest reality, as the quantity that, by virtue of moving at the finite speed of light c on the limiting boundary b of the infinite universe, possesses at the same time and in conformity with the finite-infinite equivalence principle finite and infinite magnitudes. We may also define the maximum real quantity as the quantity that has in conformity with the synthetic principle of the included third neither finite nor infinite magnitudes and hence no magnitude at all.

These definitions of maximum real quantity show us that the finite speed of light c is a gateway to infinity. In order to maximize or infinitize a finite quantity of a given kind, including the finite speed of light itself and our finite individual senses (for example, the size of our brain cells and, their rate of processing information), we need to reach the finite speed of light *c* at the limiting boundary *b* of the infinite universe. Thus, it is not the infinite quantity—that is, the infinite amount of energy required to accelerate an increasingly heavier body to the finite speed of light, which renders the finite speed of light inaccessible—but rather the other way around: it is the finite and therefore accessible speed of light that, renders the infinite amount of energy accessible. In fact, the attainment of the finite speed of light by the moving body allows the body to access the state of infinite momentum and energy. Thus light, by linking the finite quantity of a given kind with the infinite quantity of a different kind—for example, the finite speed of light with the moving body's infinite mass energy and infinite momentum—brings finitude to the infinite and infinity to the finite.[12]

This is equivalent to the process of light linking the finite part of a given quantity with its infinite part, as happens with the real speed of light taken as a complex unity that links its finite speed *c* with its contrary infinite and zero magnitudes: $c = 1 = \infty \times 0$. As the Pythagoreans clearly asserted, "The infinite must be even because grasped and limited by the odd (the finite) brings infinity to the beings."[13] In this sense, the luminous moving body behaves as the universe's outer limiting point *b*: that is, the point-fire or cosmic singularity unifying all oppositions of the Euclidean world.

In addition to the finite speed of light, we have the finite lengths 10^{26} meters and 10^{-34} meters , which, being equal to their respective infinite lengths—for example, 10^{26} m = 10^{∞} = ∞m and 10^{-34} m = $10^{-\infty}$ = 0m—simultaneously bring infinity to the finite and finitude to the infinite. In this sense, we may consider the finite lengths 10^{26} meters and 10^{-34} meters as gateways to infinity. This means that it is sufficient to reach these finite lengths located at a finite distance from us in order to immediately reach the infinite lengths 10^{∞} = ∞meters and $10^{-\infty}$ = 0meters located at an infinite distance from us. We call this process the infinitization of the finite through its equality with the infinite.

At the same time, we have the converse process, in which the infinitely distant lengths 10^{∞} = ∞m and $10^{-\infty}$ = 0m are brought at finite distances from us, which are respectively the finite lengths 10^{26} m and 10^{-34} m. We call this process the finitization of the infinite through its equality to the finite.

But how is it possible for a finite quantity of a given kind to be maximized or infinitized and, in a general manner, how is it possible for the finite to touch the infinite without being destroyed by the infinite?

The answer is that the finite cannot be destroyed by the infinite, insofar as the infinite is a complex whole divided into contrary (equal and opposite) infinites. Mutually neutralized, they allow the emergence of an intermediate finite quantity of a given kind taken as a complex indeterminate unity, which instead of being destroyed by the infinite is the very product or balanced ratio of contrary infinites—of the large infinite (τό μέγα ἄπειρον) and the small infinite (τό μικρόν ἄπειρον).

By maximally accelerating your brain from its initial lowest state *a*—that of being a simple individual in a Euclidean rest frame on earth at the center *a* and obeying analytic principles of organization (analytic ontology)—to its final highest state *b*, we discover that in its deepest reality, your individual brain is a complex infinite whole—a universe—located on the limiting boundary *b* of the infinite sphere governed by synthetic principles of organization (synthetic ontology). It follows that if you succeed in moving from *a* to *b* despite the infinite distance that separates them, it is because in its deepest reality your brain is not a finite body—a simple individual at rest or indefinitely accelerating within the Euclidean world—but rather an infinite body: that is, a universe moving with a maximum speed on the limiting boundary *b* of the infinite sphere. As an infinite body, your brain travels along the infinite distance separating *a* from *b* and counts its infinite number of parts immediately with a maximum speed, because the inner infinite distance that separates them is the outer zero distance that unites them. The maximum speed is the real speed of light *c* equal to unity, which we have defined as the product or ratio of infinite and zero speeds.

But is there any possibility of experiencing what your infinite synthetic universal reason thinks? Can you directly sense your infinite body on the infinite sphere traveling and counting the infinitely many at once or in zero time with an infinite energy and speed?

IS THE INFINITE BODY PERCEPTIBLE?

Cantor defined the infinite set "as the infinitely many that allows itself to be thought as one."
Does the infinitely many allow itself to be sensed as one?

There is a fundamental contradiction between our infinite mind, say our infinite synthetic universal reason (νοῦς), which has the power to think our infinite body rotating on the limiting boundary of the infinite sphere with a luminous speed, and our finite brain, say our individual senses at rest at the center *a* of the infinite sphere, which can neither experience directly our immanent infinite body moving at the speed of light nor sense the infinitely many as one.

This contradiction is analogous to that existing between our universal reason thinking our immanent rotation as an inhabitant of the rotating earth and our individual senses, which have no direct sensation of our terrestrial rotation or of any other kind of motion in deep space. How, then, is it possible to make uniform our finite brain with our infinite mind, our finite individual senses with our infinite universal reason, and solve this fundamental conflict?

The answer is by transforming our finite brain into an infinite brain, our Euclidean finite individual senses into non-Euclidean infinite universal senses having the power to sense the infinitely many as one, and therefore to directly experience the quality of being an infinite body computing the incomputable real 1 on the limiting boundary *b* of the infinite universe by virtue of traversing and counting its infinitely many parts at once at the maximum speed of light.

We have argued that if we employ the cosmic force of expansion (centrifugal force, or repulsive gravity) to accelerate the brain away from its rest point *a* to the limiting boundary *b* of the infinite universe, where it moves at the maximum speed of light, we transform the brain from a sensible finite body at rest or indefinitely receding from rest into a real infinite body capable of experiencing actually and immediately its proper infinite body and effectively compute the incomputable real 1. Because the brain's acceleration takes place along the Euclidean line of sight *E,* which is the horizontal axis of expansion, it is an external solution, also called the *way of acceleration*, requiring a maximum effort and time. This exoteric solution assumes that the brain is inherently at rest and therefore constantly needs an external force, for example, the cosmic expansion (centrifugal) force, the artificial machine, or the transcendent Mind, to successively accelerate it to the maximum speed of light and project the brain beyond its initial Euclidean world of rest.

Thus, given the Hubble law $v = kd$ and the Hubble constant k which is roughly 3×10^4 m/s/1.5×10^{22} m, in order to escape from the local super cluster of galaxies occurring at a distance $d = 1.5 \times 10^7$ light-years $= 1.5 \times 10^{23}$ meters, the brain needs to accelerate its speed of recession v to roughly 3×10^5m/s; to escape from the local Euclidean universe located at a distance $d = 1.5 \times 10^{10}$ light-years $= 1.5 \times 10^{26}$ meters and identified with the inner observable part of the real physical universe, the brain needs to accelerate its speed of recession v to roughly 3×10^8 m/s, which is the sensible speed of light.

However, we have another internal solution, also called the *way of rest*, in which the acceleration of the brain to the speed of light is performed through maximum mental concentration (compression) on a rest point along the vertical axis of contraction or division. It constitutes the Indian yogic method and requires a minimum effort and a minimum time—in other words, a minimum cost—because it takes place here and now, within the resting brain itself at the deepest level of its structure. This esoteric mental method assumes that the brain is inherently rotating at the maximum speed of light and has its quantities inherently maximized or infinitized; therefore, the brain needs neither an extrinsic force (cosmic centrifugal force), nor an artificial machine, nor a transcendent Mind to accelerate it to the maximum speed of light with a maximum effort or a maximum time. What the brain needs is to experience directly its immanent maximum or infinite quantities.

These complementary solutions constitute the perceptual solution to the problem of motion according to which continuous motion from *a* to *b* is necessary and real because our brain is actually and with respect to its deepest structure an infinite body that has the power to traverse and compute *ab*'s infinitely many parts in one instant at maximum speed.

NOTES

1. Concerning the properties of quantity, see Aristotle's *Organon*, trans. J. Tricot (Paris: Librairie Philosophique J. Vrin, 1984), Categories 6, 5*b*, 12.

2. As Kant clearly puts it: "If the world is a whole existing in itself, it must be either finite or infinite. But it is neither finite nor infinite—as has been shown, on the one side, by the thesis, on the other, by the antithesis. Therefore the world—the content of all phenomena—is not a whole existing in itself. It follows that phenomena are nothing, apart from our representations. And this is what we mean by transcendental ideality." See *Critique of Pure Reason*, translated by J.M.D Meiklejohn (London: Everyman's Library, 1991), Transcendental Dialectic Book II, 304.

3. See Aristotle's, *Metaphysics*, trans. J. Tricot (Paris: Librairie Philosophique J Vrin, 1986), I, 10, 1058.

4. "It even seems that from these contraries (the center and the extremity of the universe), we derive the definition of all other contraries, because the terms that have the greatest distance from each other and belong to the same whole are defined as contraries" Aristotle, *Organon*, trans. J Tricot (Paris: Librairie Philosophique J. Vrin, 1984), Categories I, 6a, 15.

5. See Aristotle's, *Metaphysics*, trans. J. Tricot (Paris: Librairie Philosophique J. Vrin, 1984), Z, 17, 1041 *b*15.

6. How do extreme opposites unite in order to form a continuous whole or universe? They unite by the geometric, mental and physical properties of the continuous universe itself. For example, we have argued that the limiting point *b* of the universe's infinite sphere, geometrically holds together the Euclidean universe's infinitely many parts regardless of their difference and distance. Operating as the mental principle of the unity or coincidence of opposites (*coincidentia oppositorum*), whose physical manifestation is the synthetic equivalence principle, the limiting point *b* unifies and reconciles all oppositions of the Euclidean universe. In a general manner, opposites that cannot be united in the low Euclidean world are united and interchanged by the limiting point *b* in the highest order of dimension zero.

The unity or coincidence of extreme opposites signifies: i) immediate contact (ἁφή), bond or link (σύνδεσμος) between opposites; ii) the composition (σύνθεσις) or mixture (μῖξις) of opposites; iii) the love or sympathy of opposites; iv) the entanglement (συμπλοκή) or fusion of opposites; v) the superposition and therefore equality of opposites; vi) the correlation of opposites: vii) the coexistence of opposites; viii) the interchange of opposites, which involves the instantaneous communication between opposites regardless of their distance. All of these meanings of the unity of opposites are expressed logically through the signs of equality (=) and intersection or conjunction (×).

7. The variation of any quantity of a given kind is inherently symmetric, taking place under the actions of the contrary (equal and opposite) forces of expansion and contraction and thus according to contrary senses—that is, according to increase and decrease. However, when we look at the variation, we perceive it as if it were asymmetric, taking place under a unique force and hence in a unique sense—that is, either under the force of expansion and hence according to increase or under the force of contraction and hence according to decrease. In this case we have a sequence of cycles of expansion and contraction under the action of a sequence of unique forces. We have here what we have already analyzed: a tension or contradiction between our individual perception of the physical whole, in which we arbitrarily reduce (contract or constraint) the physical whole into a unique determination or sequence of unique determinations, and the real physical whole existing in itself as a universe or simultaneity of determinations and hence independently of the individual observer.

8. When the body is moving at the speed of light, v/c is equal to 1 and $\sqrt{1 - v^2/c^2}$ is equal to zero. If anything divided by zero is infinite, then the $\gamma - 1/\sqrt{1 - v^2/c^2}$ factor, determining how much a quantity of a given kind varies in a moving body, is infinite with respect to the horizontal axis of expansion and its inverse, $1/\gamma = \sqrt{1 - v^2/c^2}$, is zero with respect to the vertical axis of division.

9. We can use the notations m and m' to express the fact that different masses $m \neq m'$ belonging to the same brain have different properties. Thus, m is the brain's sensible mass having at one time a unique magnitude, which we take as simple unity: $m = 1$. On the other hand, m' is the brain's real mass having at one time contrary magnitudes, namely infinite and zero magnitudes, the product of which is a complex unity: $m' = \infty \times 0 = 1$. We can also use the notations $m = 1$ and $m = \infty \times 0 = 1$ to express the fact that the same mass m has different properties. Both expressions are equivalent. The same holds true for the speed of light c. We can use the notations c, c' to express the fact that different speeds $c \neq c'$ belonging to the same light of wavelength λ have different properties. Thus, c is light's sensible speed having at one time a unique magnitude, which we take as simple unity: $c = 1$. On the other hand, c' is light's real speed having at the same time contrary magnitudes, namely infinite and zero magnitudes, that conjoined form a complex unity: $c' = \infty \times 0 = 1$. We can also use the notations $c = 1$ and $c = 1 = \infty \times 0$ to express the fact that the same speed of light c has different properties.

10. Another way of showing why breaking the light barrier has no destructive effects on the living brain is taken from the relativity theory. Because a body gets heavier the faster it moves, its energy motion is being converted into mass, according

to the celebrated equation $E = mc^2$. However, given the fact that conversion is inherently symmetric or reversible though we perceive it as if it were asymmetric or irreversible, the conversion of energy into mass implies simultaneously the converse: the conversion of mass into energy. Thus, as a body gets heavier the faster it moves, the heavier mass is converted into a greater energy motion and the body gets lighter the faster it moves. Ultimately, then, there is a circular acceleration, or a self- acceleration of the body, independently of an external propelling force, which inexorably leads the self-accelerating body to the maximum speed of light, where its energy and mass are maximized or infinitized without provoking its gravitational collapse. In fact, at the speed of light, the body has infinite heaviness and infinite lightness, which mutually neutralized produce an indeterminate and complex real body that is neither heavy nor light by virtue of being both heavy and light.

11. Based on the assumption that the finite and the infinite are equal on the limiting boundary *b* of the infinite universe, infinite mathematics considers the formulas

$$(10^{26} \times 10^{\infty} > 10^{26})_E (10^{26} \times 10^{\infty} = 10^{26})_{E'} \text{ and } (10^{\infty} \times 10^{26} > 10^{\infty})_E (10^{\infty} \times 10^{26} = 10^{\infty})_{E'}$$

to be equal. On the left side, the formula states that if we multiply the finite magnitude 10^{26} by the infinite magnitude 10^{∞}, we obtain an infinite magnitude $10^{26} \times 10^{\infty}$, which, although greater than the finite magnitude 10^{26} relative to the Euclidean line of sight *E*, is equal to the finite magnitude 10^{26} relative to the non-Euclidean line of sight *E* ′. We conclude, therefore, that the infinite magnitude 10^{∞} both increases and does not increase the finite and greatest magnitude 10^{26}, which is both variable and constant, relative and absolute.

After making the appropriate substitutions of the finite 10^{26} for its equal the infinite 10^{∞} and of the infinite 10^{∞} for its equal the finite 10^{26} in the left formula, we obtain the formula on the right side. This formula states that if we multiply the infinite magnitude 10^{∞} by the finite magnitude 10^{26}, we obtain an infinite magnitude $10^{\infty} \times 10^{26}$, which, although greater than the infinite magnitude 10^{∞} relative to the Euclidean line of sight *E*, is equal to the infinite magnitude 10^{∞} relative to the non-Euclidean line of sight *E* ′.We conclude, therefore, that the finite magnitude 10^{26} both increases and does not increase the infinite and greatest magnitude 10^{∞}, which is both variable and constant, relative and absolute.

In order to avoid the contradiction that emerges when the same finite magnitude, say 10^{26}, both varies and does not vary the infinite magnitude 10^{∞}, Cantor introduced the now famous distinction between the infinite magnitudes $10^n \times 10^{\infty}$ and $10^{\infty} \times 10^n$, which created the property of non-commutativity for the infinite magnitudes. Thus, when the finite magnitude 10^n is on the left side of the infinite magnitude 10^{∞}, the finite 10^n does not vary (increase) the infinite 10^{∞}, which is regarded as absolute or constant. We then conclude that the infinite magnitude $10^n \times 10^{\infty}$ is equal to 10^{∞}.

However, when the finite magnitude 10^n is on the right side of the infinite magnitude 10^{∞}, the finite 10^n varies (increases) the infinite 10^{∞}, which is now considered to be relative or variable. We then conclude that the infinite magnitude $10^{\infty} \times 10^n$ is greater than 10^{∞}. Now, this absolute difference and non-commutativity $10^n \times 10^{\infty} \neq 10^{\infty} \times 10^n$ is unnecessary, as we can resolve the contradiction by dividing space or observation into contrary axes or lines of sight *E* and *E*′.

12. The relativity physicists have found that the inertial mass of the moving body increases with its acceleration. Thus, the greater the moving body's acceleration rate, the heavier the body becomes. This happens because during the acceleration process, the kinetic energy of the moving body is converted, in function of the energy-mass equivalence principle, into mass. Ultimately, it would require an infinite amount of energy to resist the infinite heaviness of the moving body and reach the speed of light. However, because the energy-mass equivalence principle implies the reversible nature of the conversion process, whenever we have the conversion of kinetic energy into mass, we have simultaneously the conversion of mass into energy. Ultimately, when the moving body reaches the finite speed of light, it has both infinite mass and infinite energy, infinite heaviness and infinite lightness, which mutually neutralized leave the finite rest mass of the moving body constant despite the attainment of the speed of light.

13. See Aristotle, *Physics,* trans. H. Carteron (Paris: Les Belles Lettres, 1986), III (4), $203a_{10}$.

Chapter Five

How Powerful Is Our Brain?

How much energy can our vibrating brain possibly have? How many bits of information can our brain possibly contain and how many operations can it possibly perform in one second on its contained bits?

Let us begin with our finite brain being at rest relative to itself and vibrating (emitting/absorbing) light at any one time. Now let us assume that the brain's rest mass of 1.5 $x10^0$ kilograms is made of roughly 10 billion (10^{10}) brain cells whose radius is about 10^{-5} meters (m) and rest mass is approximately 10^{-10} kilograms (kg). Let us also assume that each vibrating brain cell stores one bit of information and emits light at any one time at the approximate rate of 10^3 cycles per second.

If we regard one cycle as one computational operation, then the brain cell, by virtue of vibrating approximately 10^3 cycles per second, performs 10^3 computational operations (ops) per second (s) or one operation in 10^{-3} seconds (in one milliseconds). This constitutes the brain cell's *rest computational power* (computational power the brain cell has at rest) operating on a time scale of the order of milliseconds. If each computational operation stores one bit, then the brain cell processes approximately 10^3 bits per second and the finite brain containing 10^{10} brain cells processes roughly $10^3 \times 10^{10} = 10^{13}$ bits of information per second, or performs roughly 10^{13} computational operations per second. This total number of computational operations expresses our finite brain's rest computational power (the computational power that the brain has when it is at rest) but also the frequency of infrared light whose wavelength is $\lambda = c/f = 3 \times 10^8/10^{13} = 3 \times 10^{-5} \approx 10^{-5}$ meters, where $\approx$ means approximately equal to. We conclude, then, that our finite brain at rest vibrates that kind of light out of the infinite spectrum of light whose wavelength corresponds to the size of its brain cells, which is approximately 10^{-5} meters and that we have called infrared light.

Infrared light of wavelength $\lambda \approx 10^{-5}$ meters and frequency $f \approx 10^{13}$cycles/second possesses an energy $E = kf$, where $k = 6.62 \times 10^{-34}$ joule seconds (Planck constant). We have therefore:

$$E = 6.62 \times 10^{-34} \times 10^{13} = 6.62 \times 10^{-21}\text{joules} \approx 10^{-20}\text{joules (J)}.$$

Because the finite brain at rest vibrates infrared light that corresponds to the size of its brain cells, we conclude that it exercises an anthropomorphic, selective or constraining effect on light.

SIMPLE BIT/COMPLEX BIT

Before computing the computational and memory powers of our brain, let us investigate the nature of the unit of information called a bit. By one bit of information we mean the smallest piece of information or knowledge stored in the smallest volume of space or in the smallest mass of the smallest vibrating body. If there is a smallest body, which we call an elementary particle, then at what level of division of matter or space will we find this elementary particle?

For the moment, let us maintain a relative conception of the elementary particle in which it is relative to a particular microcosmic scale and not relative to all scales of the microcosmos. We have, therefore: one bit stored in one vibrating particle, or one bit per vibrating particle. If relative to the biological scale the elementary particle is our vibrating brain cell, we have: one bit stored in one vibrating brain cell, or one bit per vibrating brain cell. If relative to the atomic scale the elementary particle is the vibrating electron, we have: one bit stored in one vibrating electron, or one bit per vibrating electron.

We distinguish two kinds of bits: the *simple bit*, also called the *digital bit*, and the *complex bit*, which quantum physicists call the analog or quantum bit—that is, a qubit. The simple bit is an individual that computes at one time a unique determination, that is either the alternative 0 designating, for example, the emission of a photon by the vibrating bit, or, the alternative 1 designating, for example, the absorption of a photon by the vibrating bit. We then write the following analytic proposition that verifies the analytic principle of the excluded third:

$$1 \text{ bit} = 0 + 1.$$

Here the sign of disjunction + designates *either/or*; it expresses the disjoined sum of alternatives 0 and 1. We may then read the above equation as follows: One vibrating (or rotating) bit computes at one time either 0 or 1—either emits or absorbs a photon.

We come next to the complex bit or qubit, which is a whole or universe that computes at the same time the totality of determinations: that is, both alternatives 0 and 1. We then write the following synthetic proposition that verifies the synthetic principle of the unity (or coincidence) of opposites:

1 qubit = 0×1 or 1 qubit = (01) or 1 qubit = (0 = 1).

Here the sign of conjunction ($\times$) designates *both*; it expresses the conjoined product of alternatives 0 and 1. The sign of equality (=) designates the unity of opposites 0 and 1. Both signs (=) and ($\times$) express the intimate bond existing between the opposites 0 and 1 that makes of them an undivided whole despite their division into distinct parts 0 and 1. Because of their intimate bond, the opposites 0 and 1 cannot be separated regardless of their spatial separation. We conclude, then, that they are entangled and simultaneous, that is, coexistent (or co-absent). Quantum physicists call this global simultaneity of opposites 0 and 1 the *complex superposition* of 0 and 1. In this sense, the vibrating (rotating) qubit is the complex superposition of 0 and 1, which means the qubit simultaneously computes 0 and 1—simultaneously emits and absorbs a photon.

Because the simple bit computes at one time a unique determination, that is, either 0 or 1, it implicitly raises the following simple question: Does the vibrating bit compute 0 or 1? Does the vibrating bit emit or absorb a photon? This simple question requires a simple answer and leads us to define information or knowledge as the *selection* between conflicting determinations, which we have called the contradictories (see chapter 4).

The simple bit presupposes an analytic ontology of being according to which the world is Euclidean and time-conditioned and made of discrete or simple beings (the impenetrable and isolated individuals) deprived of an intimate bond and verifying analytic principles of organization. This means that i) the fundamental laws of physics are simple and local such that different places are governed by different laws; ii) there is a simple *chronological* origin or end of the universe that generates the paradox of everything appearing from a simple nothing or destroyed into a simple nothing, that is to say, into a simple point; iii) the computational nature of the universe is digital and serial, computing at one time a *unique* determination and at different times *consecutive* determinations; and iv) the speed of light in empty space is finite and simple —namely, $c = 1$—thereby generating a time delay between near and distant parts of the universe: between here and there. It follows that due to the uniquely finite speed of light, there is no universal unity and simultaneity in the universe—there is no universal present.

Things are different with the vibrating qubit, which computes simultaneously contrary determinations—that is, both alternatives 0 and 1. In fact, the impossibility of selecting one of the qubit's alternatives, far from being a sign of our ignorance of the qubit's nature, indicates our complete knowledge of the qubit, which as a complex whole simultaneously emits and absorbs a photon. This complex and indeterminate nature of the qubit encompassing the totality of determinations leads us to define information or knowledge as the *comprehension* of coexisting alternatives, which we have called the contraries (see chapter 4).

The complex bit, or qubit, presupposes a synthetic ontology of being according to which the world is non-Euclidean and timeless and made of continuous complex beings (the interpenetrating and communicating universes), which verifies synthetic principles of organization. It follows that the fundamental laws of physics are complex universal principles, namely, principles that are the same for different parts of the universe; that there is no simple chronological origin or end of the universe: on the contrary, there is an ontological origin or end of the universe working as a complex limiting point *b* or synthetic principle unifying all isolated parts of the Euclidean plane; that as an analog (quantum) and non-serial computer, the universe *hypercomputes* simultaneously *infinitely many* alternatives, which has the effect of ending computation.

The infinite part of the speed of light ensures the instantaneous communication between near and distant parts, sensible and non-sensible parts of the universe such that there is a universal unity and simultaneity in the universe—a real universal present in which things exist by necessity, that is, according to constant universal principles, and not by accident due to external causes and varying laws.

As we argued in chapter 4, the real quantity of a given kind existing in itself and independently of the individual observer is not the simple determinate quantity produced by our individual senses but instead the complex indeterminate quantity receiving at the same time contrary determinations and verifying synthetic principles of organization. Thus, we, consider the qubit to be the fundamental and real unit of information. This means that real physical information or knowledge existing in itself and independently of individual perception is a comprehensive unit unifying opposites. On the other hand, because digital information is the effect of individual perception, which we call sensible information or knowledge, it is a selective unit disrupting the comprehensive unity of the whole—and, therefore, the coincidence of its opposites. In this sense digital or sensible information is information without comprehension.

Thus, before observation, the unit of information is the complex indeterminate bit or qubit simultaneously computing two alternatives 0 and 1. The qubit is unity in multiplicity. However, after observation, the multiplicity of alternatives alone subsists: the unity between them has disappeared. The unit of information, the complex bit or qubit, is reduced into a simple bit computing at one instant a unique determination—either 0 or 1—and at different instants consecutive determinations. The simple bit is multiplicity without unity!

Quantum physicists call this empirical disruption of the comprehensive unity or coherence of the whole by our finite individual perception, *de-coherence*. In philosophical language, we call it contradiction or conflict produced by the incomprehension of the multiple. We can express the conversion of the qubit into the bit according to the following chronological deductive order:

$$\text{qubit} = 0 \times 1 < \text{bit} = 0 + 1,$$

in which we perceive the complex qubit existing prior to our individual perception as something else—namely, as a simple bit, which is a reduced image of the original complex qubit. Thus, what we directly experience is the simple individual bit performing at one instant a unique computation, arbitrarily selected by our individual senses, and at different instants consecutive and time-consuming computations. This shows us that the purpose of deductive order is to produce a simple determinate effect from a complex indeterminate cause. Following the quantum physicists, we call the bit's simple and time-conditioned computation that obeys analytic principles of organization *serial* or *classical computation*. We call the qubit's complex and time-independent computation, which performs at the same time at least two computations and verifies synthetic principles of organization, *non-serial computation* or *quantum computation*.

Were we to perceive the world without our finite individual senses, the simple bit performing successively time-consuming computations would disappear, leaving in its place the complex qubit, which is the real unit of physical information performing simultaneously a many computations. It follows that in this work, whenever we employ the word bit, we mean the complex qubit performing simultaneously at least two computations, 0 and 1. Thus, if a vibrating body with one bit computes simultaneously 2^1= (01) alternative states of being (say emission of a photon, absorption of a photon) and a vibrating body with two coexisting (entangled) bits computes simultaneously 2^2=(00)(01)(10)(11) alternative states of being (say emission of a photon encoding a mental event, emission of a photon encoding a material event, absorption of a photon encoding a mental event, absorption of a photon encoding a material event), then a vibrating body with n coexisting (entangled) bits computes simultaneously 2^n alternative states of being.

After this digression concerning the nature of the bit, let us come back to our main topic, the exploration of our brain's maximum computational and memory powers.

THE FINITE MEASURE OF THE BRAIN'S MAXIMUM COMPUTATIONAL AND MEMORY POWERS

We have argued that when our brain is at the state of rest, it vibrates infrared light whose frequency is roughly 10^{13} cycles or operations per second. This frequency tells us the number of computational operations the brain performs in one second; in other words, it determines the brain's finite computational power at rest. Is it possible to do better than this and in conformity with the laws of the physical world? If yes, then what is the brain's ultimate and real computational power?

According to contemporary research regarding the ultimate physical limits of computation, a body's maximum limit of computational power is essentially determined by i) its total available energy, which depends on the body's mass and the finite speed of light *c*; and ii) the Planck constant *h*.[1] A body operating at the maximum limits of computational speed and memory capacity determined by the laws of physics we call the *universe* or *supreme body*. Based on Einstein's equation $E = mc^2$, the total available energy *E* of our brain with a mass of $m = 1.5 \times 10^0$ kg is

$$E = 1.5 \times 10^0 \times (3 \times 10^8)^2 = 1.35 \times 10^{17} \text{ joules (J)},$$
$$\text{or } 8.43 \times 10^{35} \text{ electronvolts (eV)}.$$

This equation tells us that if we accelerate the brain's mass at rest to the speed of light *c*, then the maximum amount of free energy we can extract from its mass at rest is $\approx 10^{17}$ joules, or $\approx 10^{36}$ electronvolts. This maximum free energy will in turn determine the vibrating brain's maximum frequency and therefore its maximum computational speed.

Based on the equation $f = E/k$, which tells us how many cycles or computational operations (ops) per second a body can possibly perform in function of its available energy, we proceed to calculate the vibrating brain's approximate maximum computational frequency or power. We then have:

$$f = E/k = 1.35 \times 10^{17} \div 6.62 \times 10^{-34} = 2.03 \times 10^{50} \approx 10^{50} \text{cycles or ops/s},$$

where f stands for computational frequency or power, E stands for energy, and k is the Planck constant 6.62×10^{-34} joule seconds.

This equation determines the greatest number of operations that our vibrating brain can possibly perform in one second when we convert its entire rest mass into free energy under extreme conditions of being.

Having determined the maximum computational power of our brain when considered as a supreme body—a universe, rotating at the speed of light—we proceed to determine the maximum number of operations it has performed since its earliest origin, some 15 billion (1.5×10^{10}) years, or $1.5 \times 3 \times 10^{17}$ seconds, ago. This earliest origin or ultimate past is in reality the universe's maximally distant limiting point *b* or cosmic singularity occurring some 15 billion light-years from our brain located here and now at the point *a*, which we take to be the center of the universe. Because we perceive through our finite retinal cells (photoreceptors) at rest only the finite part of light's speed, which is the finite speed *c* taken as unity and called the sensible speed of Euclidean space, we experience a time delay between near and distant points of the Euclidean universe. Due to this time delay, we experience the maximally distant point *b* *as if* it is the earliest (or latest) point of our present brain or universe and the maximum radius *ab* of our universe, estimated to be 15 billion light-years, *as if* it is its maximum age and history estimated to be 15 billion years.[2]

It follows that the universe's cosmic age is not a natural property of the universe itself, but rather an artificial, anthropomorphic property imposed on the physical universe by our finite individual senses at rest; in fact, they perceive the eternal physical universe *as if* it is a time-conditioned universe with a cosmic age and history, which we have called the sensible or observable universe represented geometrically by the indefinitely varying Euclidean plane.

Were we to perceive through infinite universal senses the light's real speed, which is $c = 1 = \infty \times 0$, the time delay would disappear from the universe, leaving in its place a timeless physical universe with a universal present that partially closes and stabilizes the indefinitely varying Euclidean plane.

If we multiply the brain's maximum computational power 2.03×10^{50}ops/s by the brain's maximum age or history, estimated to be 15 billion years (15×10^{9} years or $1.5 \times 3 \times 10^{17}$ seconds), we obtain the maximum number of operations performed by the brain since its earliest origin. This number of operations also computes the total computational action necessary for putting the brain together as a supreme body during its maximum lifetime:

$$2.03 \times 10^{50} \times 1.5 \times 3 \times 10^{17} = 9.13 \times 10^{67} \approx 10^{68} \text{ operations.}$$

If, however, we purify the "earliest origin" from its anthropomorphic aspect—the chronological property of earliest—what is left is an "extreme origin" that coincides with the maximally distant limiting point *b* of the limiting boundary *b*. We regard this limiting point *b* (or cosmic singularity) as i) the first principle (πρώτη ἀρχή) unifying all oppositions, ii) the ultimate containing whole embracing everything because of its greatest size, and iii) the ultimate contained part—also called the *ultimate elementary particle*, or *first particle*, embracing nothing because of its smallest size. To put this concept in ontological and epistemological terms, we regard the limiting point *b* as the ontological and epistemological principle and boundary of being and knowledge of being. If we divide the rest mass of our brain by the maximum number of operations performed by the brain during its maximum life-time, we obtain the amount of minimum available mass per operation, which is 1.5×10^0 kg $/10^{68} = 1.5 \times 10^{-68}$ kilograms, or about 10^{-68} kilograms per operation. This is the smallest of all masses belonging to the smallest of all elementary particles—the first particle—which we identified with the limiting point *b* (or cosmic singularity) taken as the ultimate contained part.

This cosmic singularity *b* with the smallest mass stores one bit of information (or knowledge) that performs one operation (vibration) per 15 billion years, or $1.5 \times 3 \times 10^{17}$seconds, which is equivalent to performing $2.2 \times 10^{-18} \approx 10^{-18}$ operations (ops) per second (s). We write, therefore: one vibrating cosmic singularity with a mass of $\approx 10^{-68}$ kilograms stores one bit or, to put it another way, one bit per cosmic singularity. This one bit per singularity performing one operation per 15 billion years computes two alternative states about the smallest thing—that is, 2^1= (01) alternatives, where 0 designates, for example, the emission of a photon by the vibrating bit/singularity and 1 designates the absorption of a photon by the vibrating bit/singularity.

If we divide the mass of the brain by the mass of the first particle, which is the cosmic singularity *b*, we obtain the number of the brain's first particles or cosmic singularities, which is approximately 10^{68}. Thus, the maximum number of operations performed by the brain during its maximum history (*temporal depth*) converted into *maximum spatial depth* is the same as the number of its first particles—that is, about 1.5×10^0 kg$/1.5 \times 10^{-68}$ kg $= 10^{68}$ cosmic singularities storing 10^{68} bits of information. This number measures the brain's maximum spatial depth, which is at the same time its *maximum memory capacity.*

Taking all this into consideration, we can say that an ultimate elementary quantity of information called a bit, or a *first bit*, is stored in the cosmic singularity *b* or first particle whose mass of $\approx 10^{-68}$ kilograms is the finite measure of the smallest of all masses. If we assume that the radius of the smallest volume in which the first particle fits is $\approx 10^{-34}$ meters, we see that the mass of the cosmic singularity storing one bit is directly proportional to the boundary area of the smallest volume, which is approximately the square of its radius. We then have the following equation:

$$1 \text{ bit per singularity} = 1\text{bit per} \approx 10^{-68}\,\text{kg} \approx (10^{-34}\,\text{m})^2 \approx 10^{-68} \text{ square meters.}$$

Based on this equation, we can say that one bit of information stored in the cosmic singularity of mass m is directly proportional to the surface area A of the smallest volume in which the singularity fits. The surface area A of the smallest volume is directly proportional to its radius R squared. We then write:

$$1 \text{ bit per } m \approx A \approx R^2.$$

Now, insofar as the first particle or cosmic singularity storing one bit of information has the finite mass $\approx 10^{-68}$kg, it is a finite, material, sensible, and measurable thing possessing extension. On the other hand, insofar as this finite mass is regarded as an ultimate or smallest mass belonging to the first particle, it is at the same time an infinite quantity according to division, that is, an infinitely small or zero mass such as $\approx 10^{-68}$ kg $= 10^{-\infty} =$ 0kg. As a zero mass, the cosmic singularity is an infinite immaterial, intelligible, and immeasurable thing—a nothing with no extension, which we identify with the *point-mind* assigning consciousness, that is, thought and sensibility, to the unconscious matter that constitutes the *quality* of matter; the ancient Ionian Greek, Persian, and Hindus thinkers of the sixth century BCE would say a *point-nous* or *point-soul* assigning intelligence and soul to the unintelligent and soulless matter.

We consider, therefore, the cosmic singularity *b* or first particle storing the first bit of information as a complex whole —a Platonic point-line (ἄτομος γραμμή), which, relative to the Euclidean horizontal axis of extension (and dimension) is a material thing with its magnitude represented geometrically by the line, whereas relative to the non-Euclidean vertical axis of division or contraction, it is a mental thing without magnitude represented geometrically by the point. This composition of opposites, analogous to the spherical *atom-soul* of Democritus, allows us to measure the immeasurable nothing of zero radius and zero mass—for example, the extensionless and massless soul or consciousness—through its measurable finite radius of $\approx 10^{-34}$ meters and finite mass of $\approx 10^{-68}$ kilograms.

This exploration of the complex nature of the first particle storing a complex bit of information leads us to conclude the rebellious Anaxagorian-based thesis, that mind or consciousness, far from being an emergent property of evolving matter, is an intrinsic property of timeless matter complementing its unconscious materiality. Paraphrasing the Pythagoreans, we assert that infinite consciousness must be even or complex, because being grasped by unconscious matter assigns a proto-consciousness to all finite material beings.

It follows that our brain, composed as it is of about 10^{68} entangled cosmic singularities or first particles that are simultaneously mental, is "an infinite and self-ruled mind" (Anaxagoras) that generates successively through its infinite number of rotations or contains simultaneously in one infinite rotation the infinite diversity of things, the infinite totality of universes.

COSMIC SINGULARITY AND THE HOLOGRAPHIC PRINCIPLE

Let us replace the concept of cosmic singularity taken as a first particle by the more general notion of the *ultimate elementary volume* or *first volume* of space, which we define as the spherical region whose radius *R* has the smallest length $\approx 10^{-34}$ meters. We affirm then, on the basis of what we have already shown, that the surface area of the volume of radius $\approx 10^{-34}$ meters determines the information content and mass of the volume and not the volume itself. This constitutes an extended form of the holographic principle of physics, which covers not only the information content of a specific volume but also its contained mass.

The holographic principle defies the empirical expectation that the capacity of a region should depend upon its volume. It also challenges our Euclidean assumption that the enveloping whole, say the 3-D volume, is greater than its enveloped part, say the 2-D surface. However, philosophically the holographic principle is perfectly justifiable.

Let us assign a positive meaning to the concept of volume, which is a region limited by a surface area, and identify it with the 3-D body. Let us now employ Aristotle's definition of body, according to which "body is that which is limited by a surface."[3] What this definition says essentially is that the 2-D limiting surface is the determining condition of the existence of the 3-D body, which is also what the extended holographic principle, in essence, says. This means that the 2-D limiting surface of the 3-D body contains the totality of information and mass included in the interior of the 3-D body.

One of the logical implications of the holographic principle is that the totality of information and mass of the containing 3-D body, or whole, occupies a 2-D surface area, which is a contained part. If such a compression of the 3-D whole into a 2-D part is possible without loss, that is, without destroying the 3-D whole, then a generalization of this principle leads us to affirm that a 3-D whole unconstrained by its three-dimensionality can occupy a 1-D line or a 0-D point without losing its immanent properties: for example, its property of three-dimensionality. An ultimate generalization of this principle affirms that an ∞-D whole has the power to occupy a 0-D point in conformity with the synthetic principle of self-containment and freedom (reflexive order, self-order, spontaneous order), which governs the universe or whole.

This principle of self-containment also explodes the Euclidean assumption that the containing ∞-D whole is greater than its contained 0-D point: 0-D point < ∞-D whole. In fact, what we have here is the converse order of things: The containing ∞-D whole is less than its contained 0-D point: ∞-D whole < 0-D point. The general property of enveloping wholes occupying enveloped parts presupposes, too, that dimensions are equivalent or continuous, allowing the continuous passage of the ∞-D whole from one dimension to another without losing its constancy in a continuous universe governed by synthetic principles of organization.

The holographic principle was first proposed by the physicist Gerard 't Hooft in 1993 and elaborated on by Leonard Susskind, who claimed that all of the information contained in a volume of space of radius *R* is directly proportional to the boundary area of the spherical volume.[4] In the beginning, the holographic principle was inspired by black-hole thermodynamics, which holds that the maximum information content of the region inside the hole is directly proportional to the area of its event horizon, more precisely to one quarter of the event horizon's area measured in Planck areas. The Planck area is the Planck length, about 10^{-35}meters squared.[5] Subsequently, it was suggested that the holographic principle has a universal validity and could be applied to all volumes of space regardless of their nature, whether they are holes or bodies.

We must distinguish the conventional concept of the black hole from our non-conventional concept of cosmic singularity. The smallest body—the vibrating cosmic singularity—is not a black hole, that is, the collapse of the vibrating body and hence a maximally entropic object. In fact, the maximum matter-energy density of the cosmic singularity, whose finite measure is roughly10^{31}grams per cubic centimeter (10^{-65} g/(10^{-32} cm)^3 = 10^{31} g/cm^3) far, from indicating the gravitational collapse of a given body, is the very source of its permanent vibration and the eternal living fire that constantly animates all bodies of the universe, ensuring thereby their permanent life.

In this sense, the cosmic singularity is a zero entropic object that, in every instant reverses and neutralizes the arrow of time. It follows that what cosmologists call black holes or singularities are in reality the ultimate constituents or first particles of the physical universe out of which everything is made.

We can express the holographic relation among information *I*, mass *m*, and surface area *A* of a specific volume of radius *R* by way of the following equations:

$$I = kA \qquad \text{and} \qquad m = kA,$$

where k is a constant of proportionality. Because the surface area A of the spherical volume is directly proportional to its radius R squared, we replace A in the above equations with R^2:

$$I = kR^2 \qquad \text{and} \qquad m = kR^2.$$

The first equation on the left side enables us to convert the surface area of the volume into information describing the volume; the second equation on the right side enables us to convert the surface area of the volume into the volume's mass.

We have here a spatial theory of information and mass according to which their origin within a specified volume is not a transcendent God, but rather space itself. In fact, as the holographic principle stipulates, it is the enveloping surface area of the specified volume that originates the information and mass included inside the volume so that volumes or physical bodies have the information and mass they do have. This means that physical bodies are not created accidentally by external causes at particular successive moments of the evolving universe. They are, instead, necessary, constant, and timeless beings existing according to constant universal principles—for example, according to the holographic principle, which governs the permanent universe.

If we take as R the length $\approx 10^{-34}$ meters, which we consider to be the smallest of all lengths, then the constant of proportionality k relative to information I is approximately: $k = I/R^2 = 1$ bit$/(10^{-34})^2$, whereas relative to mass m, it is approximately: $k = m/R^2 = 10^{-68}/(10^{-34})^2$. These constants express approximately the amount of information and mass that the boundary area of the smallest volume of space—namely, the first volume of space—can create and hold. Combining both constants, we can say that one bit of information (or knowledge) stored in one mass of $\approx 10^{-68}$ kilograms that fits within the smallest volume also occupies its surface area of $(10^{-34})^2 \approx 10^{-68}$ square meters in conformity with the synthetic principle of self-containment or reflexive order.

In line with the above mathematical principles, let us use as an example a volume of space of radius $R \approx 10^{-5}$ meters. The maximum information and mass that this spherical volume can possibly contain is approximately:

$$I = kR^2 = 1 \text{ bit}/(10^{-34})^2 \times (10^{-5})^2 = 10^{58} \text{ bits and,}$$

$$m = kR^2 = 10^{-68}/(10^{-34})^2 \times (10^{-5})^2 = 10^{-10} \text{ kilograms (kg).}$$

We will call a brain cell the body that has $\approx 10^{58}$ bits of information stored in $\approx 10^{58}$ cosmic singularities that form a mass of about $10^{58} \times 10^{-68}$ kg $= 10^{-10}$ kg, which fits inside a spherical volume of radius $R \approx 10^{-5}$ meters. If we multiply

$\approx 10^{58}$ bits by $\approx 10^{10}$ brain cells, we obtain the total amount of information that our brain can possibly have: $\approx 10^{68}$ bits of information. It is also the total amount of information required to describe and realize the brain. If we multiply $\approx 10^{-10}$ kilograms by $\approx 10^{10}$ brain cells, we obtain our brain's total mass, which is about 10^{-10} kg $\times$ $10^{10} = 10^{0}$ kg.

Another way for estimating our brain's total amount of information and mass is in terms of applying the above mathematical principles directly to the brain. Indeed, if we assume that our brain of radius $\approx 10^{-1}$ m (10cm) fits within a volume of space of radius $R \approx 10^{0}$ m, which is roughly half the height of a human being, then according to the holographic principle, our brain's total information and mass inside this specified volume are determined by the boundary area of this volume, which is directly proportional to the volume's radius squared. We have, therefore, approximately:

$$I = kR^2 = 1 \text{ bit}/(10^{-34})^2 \times (10^{0})^2 = 10^{68} \text{ bits and,}$$

$$m = kR^2 = 10^{-68}/(10^{-34})^2 \times (10^{0})^2 = 10^{0} \text{ kilograms.}$$

In function of these equations, we call the brain the body that has $\approx 10^{68}$ bits of information stored in $\approx 10^{68}$ cosmic singularities, which constitutes its maximum memory capacity. The $\approx 10^{68}$ coexisting (entangled) singularities form a mass of roughly $10^{68} \times 10^{-68}$ kg $= 10^{0}$ kg, and compute 2^n (where $n = 10^{68}$) alternative material and mental states of the brain.

Similarly, if we assume that the finite observable part of our physical universe fits within a finite volume of space of radius $R \approx 10^{26}$ meters, then according to the holographic principle, the total information and mass inside this specified volume are determined by the boundary area of the volume, which is directly proportional to the volume's radius squared. We have, therefore, approximately:

$$I = kR^2 = 1 \text{ bit}/(10^{-34})^2 \times (10^{26})^2 = 10^{120} \text{ bits and,}$$

$$m = kR^2 = 10^{-68}/(10^{-34})^2 \times (10^{26})^2 = 10^{52} \text{ kilograms.}$$

The $\approx 10^{120}$ bits of information stored in $\approx 10^{120}$ cosmic singularities constitute the observable universe's memory capacity, or, more precisely, the memory of the finite observable part of the real physical universe. This mass of roughly $n = 10^{120}$ coexisting and entangled singularities computes at the same time 2^n (where $n = 10^{120}$) alternative material and mental states of the observable universe. Thus, the maximum memory capacity of the finite observable universe composed of $\approx 10^{120}$ singularities storing $\approx 10^{120}$ bits of information is roughly 10^{52} times our brain's maximum memory capacity composed of $\approx 10^{68}$ singularities storing $\approx 10^{68}$ bits. This indicates that similar to the computational power of the body, which is directly proportional to its available energy, the memory capacity of the body is directly proportional to its mass.

THE COSMIC SINGULARITY'S INFINITE COMPUTATIONAL POWER

Having identified the ultimate elementary quantity of information—that is, the first bit of information stored in the cosmic singularity *b* of mass $\approx 10^{-68}$ kg and occupying the ultimate elementary area, about $(10^{-34})^2 = 10^{-68}$ square meters—we will proceed to determine its corresponding ultimate elementary computational power, also called its first computational power. We will show that the cosmic singularity *b*—the first particle—located at the limiting boundary *b* of the physical universe's infinite sphere vibrates *first light* at infinite and zero frequency and therefore has at the same time infinite and zero computational power.

We have argued that the cosmic singularity fits within the smallest of all volumes of space of radius $R \approx 10^{-34}$ meters and that its finite mass $m \approx 10^{-68}$ kilograms is the smallest of all masses. From this finite smallest mass, we derive the finite lowest energy available $E = mc^2 = 1.5 \times 10^{-68} \times (3\text{x } 10^8)^2 = 1.35 \times 10^{-51} \approx 10^{-51}$ joules, or $8.43 \times 10^{-33} \approx 10^{-32}$ electronvolts, and from this finite lowest energy, we derive the finite lowest temperature $T = E/k$, where k is the Boltzmann constant 8.617×10^{-5}eV/ 10^0 K. We have therefore:

$$T = E/k = 8.43 \times 10^{-33} \div 8.617 \times 10^{-5}\text{eV}/\ 10^0\ \text{K} \approx 10^{-28}\ \text{kelvins (K)}.$$

Based on the equation $f = E/k$, where k is the Planck constant 6.62×10^{-34} joule seconds, we convert the above lowest energy into the lowest frequency by means of dividing the singularity's available energy by the Planck constant, which is the minimum possible energy possessed by a vibrating body adapted to the atomic scale. We obtain, then, the singularity's finite and lowest computational frequency

$$f = 1.35 \times 10^{-51} \div 6.62 \times 10^{-34} = 2.03 \times 10^{-18} \approx 10^{-18}\ \text{cycles or operations per second,}$$

or one computational operation in $4.92 \times 10^{17} \approx 1.5 \times 3 \times 10^{17}$ seconds, or ≈ 15 billion (1.5×10^{10}) years. If we multiply the singularity's computational frequency by the number of cosmic singularities of our brain, we obtain the maximum number of computational operations our brain can possibly perform in one second on its $\approx 10^{68}$ singularities storing $\approx 10^{68}$ bits of information:

$$f = 2.03 \times 10^{-18} \times 10^{68} = 2.03 \times 10^{50} \approx 10^{50} \text{operations per second.}$$

This tells us that the brain's total computational power remains constant regardless of whether we divide it by $\approx 10^{10}$ brain cells storing $\approx 10^{10}$ bits of information or by $\approx 10^{68}$ singularities storing $\approx 10^{68}$ bits of information. As a matter of fact, the greater the number of parts by which we divide the brain's total computational power, the less is the computational power of each part. If we divide the brain's total computational power by the finite and maximum number of parts—that is, by $\approx 10^{68}$ particles—then the computational power allocated to each part is minimum: $2.03 \times 10^{-18} \approx 10^{-18}$ops/s. This finite lowest computational frequency has the finite and longest possible wavelength λ given by the following equation:

$$\lambda = c/f = 3 \times 10^{8}/2.03 \times 10^{-18} = 1.47 \times 10^{26} \approx 1.5 \times 10^{26} \text{ meters.}$$

Now we can see that the finite longest wavelength associated with the finite lowest computational frequency and energy has roughly the magnitude of the radius of the universe, or, more precisely, of the finite observable part of the physical universe. This light with the longest wavelength and the lowest frequency we will call *hyper radio*. In a general manner, by hyper radio waves we mean light whose wavelength λ lies in the range of 10^{8} to 10^{26} meters; frequency $f = c/\lambda$ lies in the range of 10^{0} to 10^{-18}cycles per second; energy $E = kf$ lies in the range of 10^{-33} to 10^{-51} joules; and temperature $T = E/k$ lies in the range of 10^{-10} to 10^{-28} kelvins.[6]

Because all quantities related to the cosmic singularity at the limiting boundary *b* of the physical universe are absolute or maximum and an absolute or maximum quantity is the finite quantity of a given kind that is at the same time infinite (according to extension or division), we define the singularity's smallest mass as the finite mass of $\approx 10^{-68}$ kilograms, which is simultaneously an infinitely small or zero mass. We stipulate, therefore, the following equality between the singularity's relative finite mass $\approx 10^{-68}$ kilograms belonging to the Euclidean series of masses and its absolute infinitely small mass $10^{-\infty} = 0$ kg transcending the Euclidean series: $\approx 10^{-68}$ kg $= 10^{-\infty} = 0$kg. Here the Euclidean finite mass of $\approx 10^{-68}$ kilograms grasps infinity according to division and brings it to the finite singularity, which seen from beyond the Euclidean series, that is to say from zero dimension, is an immaterial nothing—a point-mind, or a point-soul, of zero mass. The principle that allows the finite to grasp the infinite and bring it to the finite is, as we have argued, the cosmic singularity itself also working as a unifying principle of opposites—of the finite and the infinite, of the material and the immaterial (mental).

Thus, by virtue of the unity of the finite material and the infinite mental expressed physically in the form of the finite-infinite equivalence principle, we assign to the singularity's mental part the finite mass of $\approx 10^{-68}$ kilograms and to the singularity's material part an infinitely small or zero mass. In this way, the cosmic singularity's mental part has weight, whereas the material part is weightless and therefore ethereal.

What holds true for the singularity's finite and smallest mass, which is simultaneously an infinitely small or zero mass, also holds true for all the others quantities related to the smallest mass. Accordingly, the finite least energy $\approx 10^{-51}$ joules extracted from the finite smallest mass $\approx 10^{-68}$ kilograms and having the finite and lowest temperature $\approx 10^{-28}$ kelvins, is simultaneously equal to zero energy having zero temperature and known as the freezing point of the universe. Finally, the finite and lowest computational frequency $\approx 10^{-18}$ops/s (derived from the finite lowest energy $\approx 10^{-51}$J) that has the finite and longest wavelength $\approx 10^{26}$ meters is at the same time equal to zero computational frequency having zero energy and infinite wavelength.

To sum up, we have the following equalities between the finite and the infinite:

$$m \approx 10^{-68}\,\text{kg} = 0\text{kg};\ E \approx 10^{-51}\text{J} = 0\text{J};\ T \approx 10^{-28}\,\text{K} = 0\text{K};\ f \approx 10^{-18}\text{ops/s} = 0\text{ops/s};$$
$$\lambda \approx 10^{26}\,\text{m} = \infty\text{m}.$$

These equalities show that the finite and maximum or absolute quantities 10^{-68} kg, 10^{-51}J, 10^{-28} K, 10^{-18} ops/s, and 10^{26} m are gateways to infinity. It is sufficient to reach these limiting quantities located at a finite distance from us in order to reach their respective infinite quantities located at an infinite distance from us.

Having determined the computational power of the vibrating singularity in function of its mass energy, we will now determine its computational power in function of the size of the volume in which it fits. Let us conjecture that the first particle or cosmic singularity vibrates (emits/absorbs) light whose wavelength λ is directly proportional to the radius of the smallest volume in which it fits and which is $\approx 10^{-34}$ meters. We write, therefore, $\lambda \approx 10^{-34}$ m. From this finite and smallest wavelength, we derive the singularity's finite and highest computational frequency

$$f = c/\lambda = 3 \times 10^{8}/10^{-34} \approx 10^{42}\text{cycles or operations per second,}$$

or one computational operation in $\approx 10^{-43}$ seconds. We convert this finite and highest computational frequency into the following finite and highest energy

E = Planck constant $\times f = 6.62 \times 10^{-34} \times 10^{42} \approx 10^{9}$ joules, or 10^{28}electronvolts. This gives the following finite and highest temperature $T = E \div$ Boltzmann constant $= 10^{28} \div 8.617 \times 10^{-5}$eV/$10^{0}$ K $\approx 10^{32}$ kelvins, also known as the boiling point of the universe.

Now, insofar as we regard these finite quantities as maximum or absolute, they are at the same time infinite quantities verifying the finite-infinite equivalence principle. We write, therefore, the following equations:

$$\lambda \approx 10^{-34}\,\text{m} = 0\text{m}; \quad f \approx 10^{42}\,\text{ops/s} = \infty\text{ops/s}; \quad E \approx 10^{9}\,\text{J} = \infty\text{J}; \quad T \approx 10^{32}\,\text{K} = \infty\text{K},$$

which show that the finite and maximum or absolute quantities 10^{-34} m, 10^{42} ops/s, 10^{9} J, and 10^{32} K are gateways to infinity. It is sufficient to reach these limiting quantities located at a finite distance from us in order to access their respective infinite quantities: zero wavelength, infinite frequency, infinite energy, and infinite temperature located at an infinite distance from us.

Now, the light with the shortest wavelength and the highest frequency we will call *hyper γ-ray*. In general, by hyper γ-rays, we mean light whose wavelength λ is in the range of 10^{-16} to 10^{-34} meters; frequency $f = c/\lambda$ is in the range of 10^{24} to 10^{42}cycles per second; energy $E = kf$ is in the range of 10^{-9} to 10^{9} joules; and temperature T is in the range of 10^{14} to 10^{32} kelvins.[7]

Because there are two ways to define the computational power of the cosmic singularity storing the first bit, we regard the cosmic singularity as the composition of the highest and lowest computational powers that have, respectively, infinite and zero energy. The highest finite computational power of $\approx 10^{42}$ operations per second has infinite energy and constitutes the *boiling point* of the universe; the lowest finite computational power of $\approx 10^{-18}$operations per second has zero energy and constitutes the *freezing point* of the universe.

As the boiling point of the universe, the singularity, whose massless part we have identified with the cosmic mind or soul, is a point-fire assigning infinite energy and life to the inert and inanimate matter. The boiling property of energetic life is etymologically confirmed by the Greek word zoe (ζωή), meaning "life," and derived from the Greek verb ζεῖν, which means "to boil." On the other hand, as the freezing point of the universe, the singularity-mind-soul is a point-ice freezing all motion and assigning absolute rest to the boiling living matter. The cold property of the singularity-mind-soul is etymologically expressed by the Greek word psyche (ψυχή)—the soul—derived from the Greek ψυχρός, meaning cold.[8]

The unity and complex superposition of the opposites fire and ice realized by the singularity-mind-soul enables the latter to continuously breathe the divine fire of life to lifeless matter without burning matter itself. This is neither an incomprehensible paradox nor an inexplicable miracle, but rather the very extra-ordinary nature of the permanently living physical universe governed by synthetic principles of organization. As a matter of fact, the first natural philosophers of Ionian Greece, during the sixth century BCE—Heraclitus, for example—called this very primitive mixture and complex unity of opposites lying beneath or beyond the surface of Euclidean analytic experience the divine proportion, Logos (λόγος), or Physis (Φύσις). This divine proportion—the Logos—which we geometrically identified with the unifying limiting point *b* of the physical universe, is simultaneously the good, the beautiful, and the just and constitutes the origin, founding principle, and end of the eternally living Cosmos.

THE COSMIC SINGULARITY'S INFINITE MEMORY POWER

What is the ultimate and real memory power of the cosmic singularity, which is regarded as the ultimate elementary particle or first particle vibrating first light at zero and infinite frequencies? We have argued that the computational power of a body is determined by its available energy.

However, the body's memory power, that is, the amount of bits of information the body contains, is determined by the boundary area of the volume in which the body fits. Thus, if we take as the smallest volume of space the volume of finite radius $R \approx 10^{-34}$ meters and we fit within it the smallest finite body, which we call the ultimate constituent of matter or first particle or cosmic singularity, then according to the holographic principle, the first bit of information contained in the cosmic singularity is determined by the boundary area of the smallest volume in which the cosmic singularity fits. The boundary area is, in turn, directly proportional to the volume's radius squared.

We then stipulated the holographic principle $I = kR^2$, in which the constant of proportionality k is approximately $1/(10^{-34})^2$; it states that roughly one bit of information stored on the boundary area of about $(10^{-34})^2 = 10^{-68}$ square meters describes a volume of space (or the cosmic singularity that fits within this volume) with a radius of about 10^{-34} meters. Another way of reading the constant of proportionality $k = 1/(10^{-34})^2$ is to state that approximately one bit of information is generated by the boundary area of the smallest volume of space. This is accomplished by converting the boundary area of about 10^{-68} square meters into one bit of information.

Because the smallest finite volume of radius $R \approx 10^{-34}$ meters is simultaneously a zero volume of radius $R = 0$ meters, the total amount of information contained in the smallest volume, or in the cosmic singularity residing in the smallest volume, is both one bit of information, which describes the volume of space with a boundary area of about 10^{-68} square meters, and zero bits of information, which describes the volume of space with a boundary area of 0 square meters.

If we regard the cosmic singularity as the greatest finite body fitting within the greatest observable volume of space whose finite radius R is approximately 10^{26} meters, then in conformity with the holographic principle $I = k R^2$, where $k = 1\text{bit}/(10^{-34})^2$, the total amount of information contained in the greatest volume, or in the cosmic singularity residing in the greatest volume, is roughly 10^{120} bits of information, which describes the volume of space with a boundary area of about $(10^{26})^2 = 10^{52}$ square meters.

Now, insofar as the greatest finite volume of radius $R \approx 10^{26}$ meters is simultaneously an infinite volume of radius $R \approx \infty$ meters, the total amount of information stored in the greatest volume, or in the cosmic singularity residing in the greatest volume, is both $\approx 10^{120}$ bits of information and ∞ bits of information. The finite amount $\approx 10^{120}$ bits of information describes the volume of space with a boundary area of about 10^{52} square meters, whereas the infinite amount of bits describes the volume of space with a boundary area of ∞ square meters.

Ultimately, the cosmic singularity that fits simultaneously within the smallest and the greatest volumes of zero and infinite radii stores on the smallest finite boundary area of about 10^{-68} square meters one bit of information, defined as the product or ratio of infinite bits and zero bits. The real constant of information k is, therefore, approximately $1\text{bit}/(10^{-34})^2$, where $1\text{bit} = \infty \times 0$, or $1\text{bit} = \infty/0$.

What does this information constant tells us? It tells us that the smallest unit of information, the real first bit stored on the boundary area of about $(10^{-34})^2$ square meters, is not a simple individual having at one time a unique determination, namely one bit, but rather a complex universe: a maximum magnitude having contrary determinations, namely, both infinite bits and no bits. Thus, the real informational complexity at the deepest level of nature is not numeric or computational, that is, one bit per 10^{-68} square meters, but instead logical or non-computational, where one bit is the principle of composition or unity of opposites, of infinite bits and zero bits per 10^{-68} square meters, which the Ionian Greeks called divine Logos. As we can see, then, the real one bit is both a unit of information, which is quantitative, and the principle of the composition or unity of opposites, which is qualitative: in other words, information regarded as comprehension.

The ∞ bits of information holding 2^{∞} material and mental states of the physical universe describes the universe's infinite volume when experienced from inside and along the Euclidean line of sight E. The zero bits of information holding no material and mental states of the universe describes the universe's zero volume when experienced from outside and beyond the Euclidean line of sight E: that is, from dimension zero.

Because the entire infinite information of the infinite universe is infinitely compressed in the real first bit encoded in the cosmic singularity, it is sufficient to access this first bit stored on the boundary area of the smallest finite volume of radius $R \approx 10^{-34}$ meters and hence located at a finite distance from us in order to immediately acquire infinite information or memory about the infinite volume of the universe according to increase and decrease.

"To see the world in a grain of sand…to hold infinity in the palm of your hand and eternity in an hour" is a composition of opposites, as imagined by the seventeenth–eighteenth century English poet William Blake: neither a paradox, nor a magical thought, nor a poetic metaphor, but instead the very logical necessity of the self-containing ∞-extended and ∞-dimensional universe registered in its limiting point *b* regarded as the first particle or cosmic singularity residing in the smallest volume.

As a self-containing whole, the cosmic singularity is both the ultimate contained part and the ultimate containing whole. This synthetic unity of opposites, called by the ancients divine Logos or Physis, is the product of the Euclidean positive order, in which the ultimate part is less than the ultimate whole, and the non-Euclidean negative order, in which the ultimate whole is less than the ultimate part. Thus, according to this negative order, the infinite bits of information describing the infinite volume of the physical universe are contained in one bit of information occupying the smallest finite boundary area of about $(10^{-34})^2$ square meters, which is simultaneously a boundary area of zero square meters: $(10^{-34})^2\,m^2 = 0m^2$. In other words, the infinite volume of the universe is not only compressed on the boundary area of the smallest finite volume in which the cosmic singularity fits, but is also compressed on the boundary area of zero volume; and all this is accomplished without the infinite compression destroying the ∞-extended and ∞-dimensional volume of the physical universe.

Now, the infinite bits of information about the physical universe, which are compressed within the first bit, can be successively unfolded to us if we perceive the first bit as a Turing machine computing these infinite bits successively on the Euclidean plane; or they can be instantaneously revealed to us if we think of the first bit as an extra-ordinary Turing machine—a "supreme being"— hypercomputing all these infinite bits simultaneously on the infinite sphere.

THE INFINITE MEASURE OF THE BRAIN'S MAXIMUM COMPUTATIONAL AND MEMORY POWERS

If each cosmic singularity is the composition of highest and lowest computational power having respectively infinite and zero energy, then insofar as our brain is made up of cosmic singularities, its real computational power is not the finite maximum computational power of $\approx 10^{50}$ operations per second telling us how big the brain's computational power can possibly become. Rather the real computational power is the brain's finite and maximum computational frequency $f \approx 10^{50}$ops/s, which we take as a complex indeterminate unity and define as the product or ratio of contrary frequencies having infinite and zero magnitudes:

$$f \approx 10^{50}\text{ops/s} = 1 = \infty \times 0, \qquad f \approx 10^{50}\text{ops/s} = 1 = \infty/0.$$

In other words, the brain's real computational power is its capacity to accomplish the greatest computational action with the least computational action in agreement with the physical laws of greatest and least actions. This constitutes the real complexity of the brain's computational power, which is logical or mathematical, in the sense of establishing a proportion between the different parts (magnitudes) of the real computational power, and not numeric selecting one of the alternatives.

This synthetic definition of the brain's real computational power shows us that $\approx 10^{50}$ operations per second is a gateway to infinite computational power according to increase and decrease. It answers the question of how many finite operations our finite brain must perform to be an infinite brain performing an infinite number of operations in one second. Indeed, it is sufficient to maximize the brain's rest computational power of about 10^{13}operations per second either by accelerating its rest mass to the maximum speed of light or by accessing its inner cosmic singularities while being at rest. In either way, we obtain a maximum computational power, which is both finite, namely 10^{50}operations per second, and infinite (according to increase and decrease).

As a matter of fact, when the brain's rest mass of $\approx 10^{0}$ kilogram is accelerated to the finite and highest speed of light c, we can extract from this maximally accelerated mass a maximum energy that is both the finite mass energy of $mc^2 \approx 10^{17}$ joules and the infinite mass energies of infinite and zero joules upon which the brain's infinite computational powers depend (see chapter 4). Thus, when we convert the brain's rest mass into free energy, what we essentially unlock is the *infinite energy* of its finite rest mass made up of cosmic singularities. This process of unlocking infinite energy is accomplished symmetrically, according to increase and decrease, in order to conserve the stability of the brain.

What holds true for the brain holds true for any specified volume of space or body made up of cosmic singularities. For example, let us take a specific body whose contained mass is roughly 10^{52} times the mass of our brain and which we call the universe, or, more precisely, the finite observable part of the physical universe. We say that the observable universe's real computational power is not how finitely big it actually is, namely $\approx 10^{52}$ times our brain's finite and maximum computational power; the real computational power is the observable universe's finite and maximum computational frequency $f \approx 10^{50} \times 10^{52} \approx 10^{102}$ ops/s, which we take as a complex indeterminate unity and define as the product or ratio of infinite and zero computational frequencies: $f \approx 10^{102}$ops/s $=1= \infty/0$.

If the first bit stored in every cosmic singularity is a real bit—a complex universe whose real memory power is not one bit but rather the product or ratio of infinite and zero bits per smallest boundary area—it follows that the real memory power of our brain made roughly of 10^{68} cosmic singularities is 10^{68} times the real one bit: $10^{68} \times 1$bit $= \infty \times 0$ or 10^{68} bits $= \infty/0 = 1$. This shows us that the real memory power of our brain is not 10^{68} bits telling us how big the brain's memory power can possibly become; the real memory power, is our brain's finite and maximum memory power of 10^{68} bits, which we take as a complex indeterminate unity and define as the product or ratio of contrary memory powers having infinite and zero bits.

This synthetic definition of maximum memory power shows us that the brain's finite and greatest memory capacity 10^{68} bits $= \infty/0 = 1$ is a gateway to infinite memory capacity according to increase and decrease. It answers the question of how many finite bits of information we need in order to obtain an infinite brain with infinite computational and memory powers counting and storing its ∞ number of available bits that register 2^{∞} material and mental states about the brain in zero time and with zero computational action and memory. In fact, it is sufficient to maximize the brain's finite memory of about 10^{22} bits by accelerating its mass to the speed of light or by accessing its ultimate constituents—its cosmic singularities—while being at rest in order to obtain an infinite memory storing an infinite number of bits that hold an infinity of alternatives at once, that is, in zero time.[9] Because zero time is the end of time, we conclude that it is impossible to obtain an infinite memory without implying the end of time and hence the end of computation depending upon time.

Based on the above, we conclude that the real complexity of the brain's memory is not determined by counting the finite number of bits contained within the brain and which constitutes the *numeric* or *computational complexity* of memory power; the real complexity of memory is determined by the proportion existing between different memory powers, or between different parts (magnitudes) of the same memory power; it constitutes the memory's non-computational or logical complexity. It is precisely this *non-computational* or *logical complexity* that ends the Euclidean analytic notion of quantity according to which no quantity admits contrary magnitudes, and hence no quantity is complex and indeterminate.

We took great pains in this work to show the opposite—that the real quantity is the Zenonian maximum quantity, which replicates the real physical universe having the power to receive contraries, namely finite and infinite magnitudes at the same time, to which we correspond the real number 1. We located the real quantity on the limiting boundary *b* of the physical universe, which is the topos where the sensible finite quantity of a given kind admitting at one time a unique magnitude becomes the Zenonian real infinite quantity admitting at one time contrary magnitudes.

Taking into consideration the above, we call the *supreme brain*, or *infinite universal brain*, the brain whose sensible finite quantities at rest—for example, its rest computational power of $\approx 10^{13}$ops/s and its total finite memory power of $\approx 10^{22}$ bits—have reached the maximum point *b* allowed by the laws of physics, at which point they become, respectively, the real infinite quantities $\approx 10^{50}$ ops/s and $\approx 10^{68}$ bits taken as an indeterminate unity and resolved into infinite and zero computational and memory powers. We have claimed that this maximum or limiting point *b* is located on the limiting boundary *b* of the universe's infinite sphere, or, more precisely, at the outer limiting point *b* of the limiting boundary *b* of the universe occurring at zero dimension and at a right angle to the Euclidean line of sight *E*.

A maximally distant observer located at 1.5×10^{10} (15 billion) light-years away from us and hence at the maximally distant point *b* of the limiting boundary *b* will observe with a perfect telescope our finite brain at rest here and now at the center *a* as an infinite universal brain rotating at a luminous speed and having infinite computational and memory powers whose finite measures are respectively $\approx 10^{50}$ ops/s and $\approx 10^{68}$ bits of memory. On the other hand, our finite brain at rest here and now at the center *a* has no immediate sensation of its immanent rotational motion at a luminous speed and of its infinite computational and memory powers. In fact, what the finite brain immediately senses during its lifetime is its state of rest, where its finite computational and memory powers are, respectively, $\approx 10^{13}$ ops/s and $\approx 10^{22}$ bits.

However, through its perfect telescope, the finite brain at rest at the center *a* observes the maximally distant observer at the maximally distant point *b* as an infinite universal brain having infinite computational and memory powers and rotating at a luminous speed. Ultimately, each observer perceives oneself as a finite brain at rest at the center *a* and the maximally distant observer as an infinite brain rotating at the speed of light on the limiting boundary *b* of the infinite sphere, and has no immediate perception of her or his immanent infinity and rotation at a luminous speed.

If each observer is a finite individual brain at rest at the center *a* of the infinite universe and an infinite universal brain rotating at the speed of light on the limiting boundary *b* of the infinite universe, we may geometrically represent this picture of the infinite universe as two infinite spheres *u* and *u′* centered on the two maximally distant observers and- intersecting in such a way that the center *a* of each infinite sphere lies on the limiting boundary *b* of the other infinite sphere. These two intersecting and communicating infinite spheres can be equally regarded as two interpenetrating spherical waves that fit around the two centers *a* (see Fig. 5.1).

Empirical science regards the *physical limit* as a place where a being disappears or collapses. For example, according to conventional science, something the maximum speed of light tells us is how much a finite body can accelerate before it collapses into a singularity of infinite curvature, infinite density, and zero extension. Here the hidden infinity linked to the finite speed of light indicates the inaccessible and lethal nature of the luminous speed. We have argued, however, that this is not true. It is not the infinite that renders the finite speed of light inaccessible and lethal, but rather the opposite. It is through the finite speed of light, which is an accessible speed because it is finite, that the finite material body can access the infinite. This infinite, far from destroying the finite (Aristotelian thesis), coexists with the finite (Ionian Greek and Hindu thesis) in conformity with the finite-infinite equivalence principle, which governs the complex physical universe and all complex maximum or real quantities related to the physical universe such as the real speed of light $c = 1$ resolved into infinite and zero speeds.

It follows that for the metaphysician, the physical limit, far from being a place where the indefinitely accelerating finite brain a_n collapses, it is a place where a_n can be *completed*, becoming an infinite brain by virtue of accessing the condition of its possibility, which is the real limiting point *b*. To put it in computational terms, the physical limit is the physical universe's accessible limiting boundary *b* where the indefinitely computing finite brain a_n becomes an infinite brain equivalent to the infinite mind that computes *maximally* (or hyper-computes) up to the real infinite whole 1 after having counted in one instant its stored infinite number of bits.

Throughout this work, we have exposed the different meanings of the real limiting point *b* of the unlimited series of parts, which are i) the real limiting boundary *b* of the physical universe defined as the sum total of its infinite number of parts and numbered by the real infinite whole1; ii) the cosmic singularity of infinite curvature enclosing the Euclidean space-time of zero curvature; iii) the point-fire that radiates light at infinite and zero frequency and continuously assigns energy and life to inert and lifeless matter; iv) the

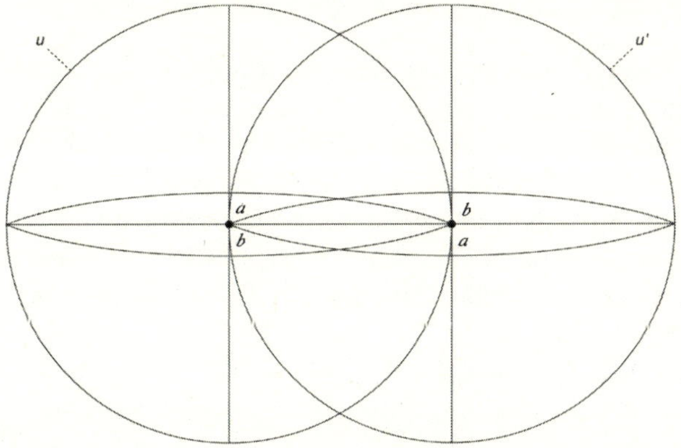

Figure 5.1. Two infinite spheres *u* and *u´* of the same radius *ab* intersect in such a way that the center *a* of each infinite sphere lies on the limiting boundary *b* of the other infinite sphere. Each observer relative to herself is a finite brain at the center *a* of her proper infinite sphere and relative to the maximally distant observer is an infinite brain on the limiting boundary *b* of the other infinite sphere. Thus, each observer is a complex whole admitting contrary determinations simultaneously.

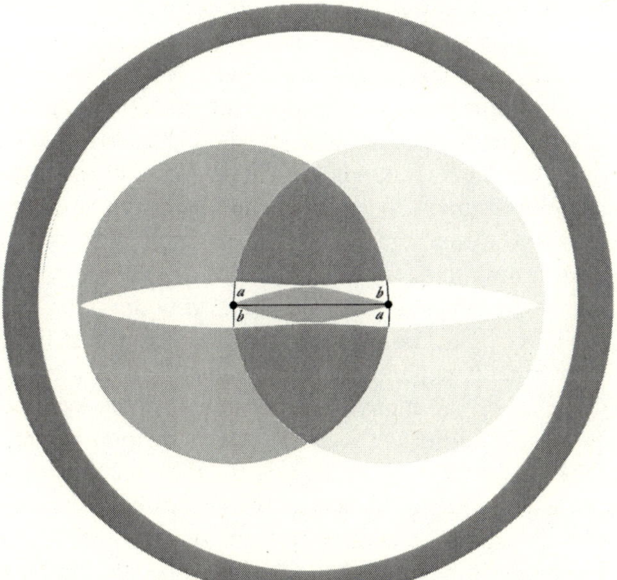

Figure 5.2. Artistic expression of the intersection of two infinite spheres where the center *a* of each infinite sphere lies on the limiting boundary *b* of the other infinite sphere. Realization: Setareh Korkchi.

infinite self-ruled mind or soul that assigns form and consciousness to the formless and unconscious matter; v) the point-logos or first principle of the unity of opposites that unifies all oppositions of the Euclidean world; vi) the first origin, final end, and ultimate constituent of all things, which metaphysicians identify with the Divine Being, epistemologists identify with the real thing transcending or underlying all phenomena, mathematicians identify with the real number 1, physicists with the first particle, and finally we identified with the real bit.

If the physical limit is the real limiting boundary *b* of the infinite sphere of the physical universe, the place where something becomes something else or is in contact with something else, then we may consider it as the place where the finite brain (the individual) becomes an infinite brain (a universe) after having accessed its inner infinite power source, namely, its inner cosmic singularity radiating infinite light that carries infinite information about the real physical infinite whole encoded in the cosmic singularity—in the real bit.

It is this divine infinite power and information source that makes the brain infinite in life, motion, and perception, as well as in computational and memory powers. For example, we have infinite perception, when the observer transcends the indefinitely extending (and dividing) Euclidean plane a_n of the sensible or phenomenal world and experiences directly the whole of the unlimited Euclidean plane: that is, the real physical world *b* numbered by the real 1. Similarly, we have infinite perception, when the perceiver transcends the neural phenomena of the subject's brain and experiences directly the subject's real mind—its real infinite thoughts and feelings. If a_n, which we identify with the starting point *a*, stands for the neural phenomena of the subject's finite brain and *b* stands for the subject's infinite mind or real self, then the problem of moving beyond the subject's neural phenomena in order to attain direct knowledge of the subject's real self is the ancient Greek problem of motion from *a* to *b*.

As the ultimate origin and principle of the unity of opposites of the physical universe, the unifying cosmic singularity *b* becomes in its logical and physical form the synthetic principle of equivalence, which stipulates the equality of opposites. One of the many manifestations of the synthetic equivalence principle is the finite-infinite equivalence principle, which stipulates the equality of the finite and the infinite. It is an omnipotent principle on behalf of which we realize all impossible tasks, such as moving from *a* to *b* after having travelled in one instant the actually infinite distance and its actually infinite number of parts lying between the extremes *a* and *b*. This instantaneous action at a distance is accomplished by passing simultaneously through the inner Euclidean infinite distance and beyond the inner infinite distance through the outer convex part of the limiting boundary *b*, which we identified with the simple singularity *b* of zero distance. The path unifying the inner infinite distance with the outer zero distance is the complex limiting boundary *b*, or the complex singularity *b* resolved into infinite and zero distances.

We have argued that if we are to accomplish a seemingly impossible task such as computing the real limiting point *b* by moving from *a* to *b*, we must first travel and compute in one second the actually infinite number of parts lying between *a* and *b*. This means that our finite brain at rest on the Euclidean plane is in its deepest and highest reality according to division and extension an infinite brain on the limiting boundary *b* of the infinite universe rotating at the speed of light and having the power to travel and compute simultaneously the actually infinite number of parts that lie between *a* and *b*. As a finite brain at rest here and now, however, we have no immediate perception of our immanent infinite computational and memory powers, which are indirectly observed (or observed with a time delay) by a maximally distant observer located 15 billion light-years away from us.

At this point, two fundamental questions emerge: First, according to what principle can our finite brain benefit from its immanent infinite computational and memory powers involving extreme energies and temperatures without being frozen at zero degrees kelvin or burned at infinite degrees kelvin? In other words, according to what principle can we have an infinite brain with infinite power at room temperature and in a controlled manner here and now at the center *a*? We leave this question open. It can be easily answered by the reader, at least in principle, after a careful reflection on the philosophical definition of maximum quantity and on the complex idea of infinity developed in this work.

Second, what are the practical modalities for obtaining here and now a stable infinite brain at room temperature? We know the finite and maximum amount of information we must have, namely 10^{68} bits of information encoding 2^{68} alternative states, to become an infinite brain. But we do not know how to access this finite amount of information stored in 10^{68} cosmic singularities and constituting a gateway to infinity with respect to computation, memory, perception, and motion. For example, we do not know how to detect light emitted from each cosmic singularity at the lowest and highest frequencies, at the longest and shortest wavelengths carrying information or memory about the infinite universe registered in each cosmic singularity, in each real bit. If we knew how to detect these emitted frequencies and wavelengths, we would have obtained an infinite universal perception of the infinitely distant and the infinitely many at once. We would have then completed Cantor's definition of the infinite set. Thus, as an infinite brain, we would have defined the infinite set as the infinitely many that allows itself to be thought and *sensed* as one! In this way, we would have closed the circle reconciling our finite sensations with our infinite thoughts, our finite brain with our infinite mind.

The ancient thinkers of Ionia and India developed different meditation techniques for unlocking our brain's infinite energy, but these are not known to us. According to some Hindu sacred texts (the Upanishads), mental concentration on a point unlocks the brain's infinite electromagnetic energy of infinite and zero temperatures. Distributed in the form of temperate neural (electrochemical) energy throughout the human body, it assigns to the human body the power of universal communication and infinite sensibility, such as sensing the infinitely distant and the infinitely many instantaneously. The ultimate purpose of obtaining the power of universal communication is to emancipate the finite brain from its incomplete time-conditioned existence. Trapped as it is with insoluble contradictions and paradoxes that obstruct the access to its own infinite power source, its inner cosmic singularity, the finite brain can generate only illness, isolation, and death. We call spiritual enlightenment the emancipation of the finite brain from irreversible time considered as the source of all contradictions and evils.

However, we have neither a theoretical nor a practical knowledge of how these meditation techniques effectively unlock and harness the sacred infinite fire of our finite brain in a natural, effortless, and controlled manner. For the moment, then, our subject belongs to the domain of speculative metaphysics. It will become a subject of experimental metaphysics as soon as empirical science realizes that the shortest path to the distant infinity is not the rectilinear path situating infinity *externally* to our finite brain and inside the artificial computer, but rather the circular path where the extreme opposites, the finite and the infinite, coincide here and now within our proper brain regardless of their infinite separation.

WHY IN THE FUTURE WE DO NOT NEED COMPUTERS

> *Anything which is coming into being is incomplete and in progress toward its principle.*
>
> —Aristotle, *Physics*

If our finite brain is in its deepest and highest reality an infinite brain deriving infinite energy, life, and information from its own infinite power source—its inner cosmic singularity—then what is the point of building computers that indefinitely extend our finite brain and life?

The dominant idea among technologists that artificial machines can do better than human beings is grounded in Aristotle's finitist assumption that there is no proportion between the finite and the infinite, that nothing finite is infinite and nothing infinite is finite.[10] Thus, if there is an analytic separation between the finite and the infinite, if no finite brain is infinite, then it is necessary to build potentially infinite artificial machines if we want to indefinitely extend our finite brain's power. Because finitism produces unsolvable contradictions and conflicts—for example, the conflicting divide between the actually finite brain and the potentially infinite artifact that causes its proper collapse—it cannot constitute a true theory of the physical world and cannot assign a meaningful end to our accelerating growth. In fact, it is an epistemological and societal dystopia. It follows that only its negation, infinitism—the Ionian Greek and Hindu theory of the sixth to fifth centuries BCE asserting that there is a proportion between the finite and the infinite—can constitute a true theory of the physical world that liberates us from the conflicting analytic divide between finite brain and infinite artifact and assigns to our accelerating growth an intelligible end.

We have argued that infinite energy is within our brain's finite mass when seen in its highest and deepest reality as a maximally accelerated brain of infinite mass and zero volume and hence of infinite density, which we called the real limiting point *b* or cosmic singularity radiating light at zero and infinite frequencies and residing outside and inside our brain. This led us to affirm that the future of humankind is not the indefinite artificial extension of our finite brain, but rather the end of artificial extension and the direct experience of our inner singularity's infinite energy and information (memory), those qualities that make us an infinite brain moving at the speed of light and naturally possessing infinite powers of computation and memory.

Seen from this perspective, the meaning of our accelerating growth is not to build artifacts that will falsely extend and eventually replace our increasingly dependent brain. It is in fact the other way around: to build artifacts that will help us to understand the way our finite brain functions at rest on earth within its Euclidean rest frame. These artifacts, however, cannot help us to understand the way our brain functions at extreme conditions of maximum acceleration or mental concentration at its deepest and highest reality on the cosmological and quantum limiting boundary *b* of the physical universe where it is an infinite mind or brain. Indeed, at this limiting boundary condition, the brain functions according to the synthetic principle of the unity of opposites known as the divine Logos or Physis, and hence in a complex, incomputable manner, which is alien to the artifact employing analytic computational logic.

It is precisely this fatal confrontation between artificial analytic logic and natural synthetic logic at the exotic limiting boundary *b* of the physical universe that will provoke the inexorable collapse of the artifact together with its finite analytic paradigm. The end of chimeric artifacts will bring about the end of computation employing Euclidean time and finite analytic logic and the triumph of continuous infinite Nature over the discontinuous computing number. In fact, both computation selecting between conflicting alternatives and Euclidean time using external causality to generate, explain, predict, or prove things are obsolete and useless, if, benefiting from our inner infinite power source, we have infinite computational and memory powers to compute and store an infinite number of alternatives at once.

So where is the real action? It lies within our brain's inner cosmic singularity of infinite density, from which we can safely extract infinite energy and information with zero cost and at room temperature. We call the knowledge for accomplishing such an extraordinary task by natural means *natural engineering of the Infinite in Act*. It is the new intelligible frontier of our accelerating growth.

NOTES

1. For a scientific overview of the question of physical limits to computation by a specialist, see "*Ultimate Physical Limits to Computation*" (arXiv: quant-ph/9908043v3 14, February 2000), an excellent article of Seth Lloyd.

2. Each retinal cell (photoreceptor) whose radius is roughly 10^{-7} meters detects that finite part of light, called optical light, whose wavelength λ is roughly the size of the retinal cell and whose frequency f is about 3×10^{15}cycles per second. It follows that the light's observable speed in empty space is the finite speed: $c = \lambda \times f = 10^{-7} \times 3 \times 10^{15} = 3 \times 10^{8}$ meters/second, which we take as unity. Thus, other things being equal, the observable finite speed of light is dependent on the observable wavelength, which is inversely proportional to the light's observable speed. Conversely, the observable wavelength of light is directly proportional to the size of our retinal cell at rest. However, if the size of our retinal cell shrinks to zero, as is the case when the brain moves at the maximum speed of light, then the observable wavelength of light would be zero and its corresponding observable frequency and speed, which are inversely proportional to wavelength, would be infinite. In this case, because of the observable infinite frequency or speed of light the time delay between here and there, near and distant parts of the universe, would have been abolished and there would have been an absolute and universal present unifying the infinite multiplicity of the universe at once.

3. Aristotle, *Physics*, trans. H. Carteron (Paris: Les Belles Lettres, 1986), III (5) 204 b_4.

4. See also Gerard 't Hooft, "*The Holographic Principle*" (arXiv:hep-th/0003004v2, 1- May 2000) and Leonard Susskind, "*The World as Hologram*"(Lecture, 2000).

5. The standard convention is to consider the Planck length, about 10^{-35} meters, as the ultimate elementary length or first length. For reasons of symmetry, however, we take in this work the first length to be about 10^{-34} meters (see chapter 4). Thus, instead of taking as the ultimate elementary boundary area or first area to be the Planck area, about $(10^{-35})^2 = 10^{-70}$ square meters, we take the first area to be approximately $(10^{-34})^2 = 10^{-68}$ square meters.

6. The calculation of the frequency f of hyper-radio waves is accomplished on the basis of the equation $f = c/\lambda$, where the speed of light c is 3×10^8 meters per second. Thus, for wavelength $\lambda \approx 10^8$ meters, we have the computational frequency

$$f = 3 \times 10^8 / 10^8 \approx 10^0 \text{ cycles per second,}$$

and for wavelength $\lambda \approx 10^{26}$ meters, we have the computational frequency

$$f = 3 \times 10^8 / 10^{26} \approx 10^{-18} \text{ cycles per second.}$$

These equations consider the observable speed of light c to be constant and the wavelength of light variable. We obtain the same results if we take the opposite direction and consider the observable speed of light $c = 3 \times 10^8$ meters/s variable and the wavelength of light $\lambda \approx 10^{-7}$meters constant.

On the other hand, based on the equation $E = kf$, where k is Planck's constant 6.62×10^{-34} joule seconds, we can calculate the energy of the hyper radio waves. For frequency $f \approx 10^0$ cycle per second, we have the energy

$$E = 6.62 \times 10^{-34} \times 10^0 \approx 10^{-33} \text{joules, or} \approx 10^{-14} \text{electronvolts,}$$

and for frequency $f \approx 10^{-18}$ cycles per second, we have the energy

$$E = 6.62 \times 10^{-34} \times 10^{-18} \approx 10^{-51} \text{ joules, or} \approx 10^{-32} \text{electronvolts.}$$

Finally, based on the equation $T = E/k$, where k is Boltzmann's constant 8.617×10^{-5} eV/10^0 K, we estimate the temperature of the hyper radio waves. For energy $E \approx 10^{-14}$ electronvolts, we have the temperature

$$T = 10^{-14} \div 8.617 \times 10^{-5} / 10^0 \approx 10^{-10} \text{ kelvins,}$$

and for energy $E \approx 10^{-32}$ electronvolts, we have the temperature

$$T = 10^{-32} \div 8.617 \times 10^{-5} / 10^0 \approx 10^{-28} \text{ kelvins.}$$

7. The calculation of the frequency f of hyper γ-rays is accomplished on the basis of the equation $f = c/\lambda$, where the observable speed of light c is 3×10^8 meters per second. Thus, for $\lambda \approx 10^{-16}$ meters, we have the computational frequency

$$f = 3 \times 10^8 / 10^{-16} \approx 10^{24} \text{ cycles per second,}$$

and for $\lambda \approx 10^{-34}$ meters, we have the computational frequency

$$f = 3 \times 10^{8} / 10^{-34} \approx 10^{42} \text{ cycles per second.}$$

Based on the equation $E = kf$, where k is the Planck constant 6.62×10^{-34} joule seconds, we can convert the frequency of the hyper γ- rays into energy. For frequency $f \approx 10^{24}$ cycles per second, we have the energy

$$E = 6.62 \times 10^{-34} \times 10^{24} \approx 10^{-9} \text{ joules, or} \approx 10^{10} \text{ electronvolts,}$$

and for $f \approx 10^{42}$ cycles per second, we have the energy

$$E = 6.62 \times 10^{-34} \times 10^{42} \approx 10^{9} \text{ joules, or} \approx 10^{28} \text{ electronvolts.}$$

Finally, based on the equation $T = E/k$, where k is the Boltzmann constant 8.617×10^{-5} eV/10^{0} K, we can estimate the temperature of the hyper γ-rays. For energy $E \approx 10^{10}$ electronvolts, we have the temperature

$$T = 10^{10} \div 8.617 \times 10^{-5}/10^{0} \approx 10^{14} \text{ kelvins,}$$

and for $E \approx 10^{28}$ electronvolts we have the temperature

$$T = 10^{28} \div 8.617 \times 10^{-5}/10^{0} \approx 10^{32} \text{ kelvins.}$$

8. For a brief discussion of the ancient doctrines concerning the soul as expounded by the first natural philosophers of Ionian Greece (Thales, Heraclitus, Anaxagoras), see Aristotle, *On the Soul*, trans. J. Tricot (Paris: Librairie Philosophique J. Vrin, 1992), I, 2.

9. If one vibrating brain cell at rest performs $\approx 10^{3}$operations per second, then during a finite lifetime of 10^{2} years $= 3 \times 10^{9} \approx 10^{9}$s, the one finite brain cell performs $10^{3} \times 10^{9} \approx 10^{12}$ operations. If one operation stores one bit, then the total memory power of the finite brain cell during one hundred years is $\approx 10^{12}$ bits. It follows that the total memory power that our finite brain at rest has during its hundred years is 10^{12} bits $\times$ 10^{10} brain cells $\approx 10^{22}$ bits. The ratio of the finite brain's memory at rest to its processing speed is the time necessary for the brain to run through its memory once. This time is approximately: $10^{22}/10^{13}$ops/s $=10^{9}$s. Now, if our brain moving at the maximum speed of light performs $\approx 10^{50}$operations per second, then during a maximum lifetime of 15×10^{9} years (15 billion of years) our brain performs $\approx 10^{68}$ ($2.03 \times 10^{50} \times 1.5 \times 3 \times 10^{17}$) operations. If one operation stores one bit, then the maximum memory power of the brain is $\approx 10^{68}$ bits, which taken as a complex indeterminate unity, we consider as the brain's real memory power resolved into infinite and zero memory powers.

10. Aristotle, *On the Heavens*, trans. J. Tricot (Paris: Librairie Philosophique J. Vrin, 1986), I, 5-7.

Chapter Six

Toward an Accomplished Humanism

A Critique of the Finite Analytic Paradigm of Contemporary Empirical Science

And Nature is but the name that men give to the mixture.

—*Empedocles*

THE OBSERVABLE UNIVERSE IS SIMPLE, EUCLIDEAN, AND TIME-CONDITIONED

According to the finite analytic paradigm of empirical science, the sensible or observable universe is assumed to be simple, Euclidean, and time-conditioned, and made of discrete or simple parts—the individuals— arranged in succession.

This sequence of individuals forms a collection of isolated individuals with no unifying bond among them and governed by analytic principles of organization. The main analytic principles of organization are self-identity, contradiction, excluded third, inequality and temporal order (heteronomous order). This last principle organizes the individuals hierarchically according to subordination and comparison.

Another assumption of empirical science is that the sensible speed of light in empty space is simple, finite, namely *c*, and equal to unity. The finite speed of light generates time delay between near and distant parts of the universe, between here, which is called the present, and there, which is called, depending on whether light is received from there or emitted to there, the past or the future. Because of the absolute separation between here and there, now and then, the universe is non-uniform, such that different parts have different properties and laws.

It follows that in a non-uniform universe, the laws are simple and local, and that there are no complex universal laws unifying different parts of the universe. Empirical scientists suffer from a double illusion: first, *perceptual illusion*, which occurs when they take what they experience through their finite individual senses, namely the Euclidean and time-conditioned universe, which we have called the sensible or observable universe *as if* it were the real physical universe and the simple finite speed of light generating time-delay between here and there, now and then, receptor and emitter, *as if* it were the real speed of light; and, second, *epistemological illusion*, which occurs when they take the analytic principles organizing the finite observable part of the physical universe, *as if* they were the real true and objective (observer -independent) principles of the physical universe. Thus, external causality (or heteronomous causality), depending on temporal order and stipulating that everything that happens has its cause outside itself, is considered together with the analytic principle of inequality and temporal order *as if* it were the real principle of the physical universe.

THE PARADOX OF THE INDEFINITELY ACCELERATING OBSERVABLE UNIVERSE

Because the observable universe is conditioned by linear (irreversible) time, it is a dynamic, indefinitely accelerating universe, which, under the action of a unique force, indefinitely recedes from its initial point $a = 0$ at the center a and indefinitely approaches its final point $b = 1$ on the physical universe's limiting boundary b. This indefinitely accelerating observable universe can be represented by the finite variable a_n, to which our individual senses apply the arrow of time working either as a force of repulsive gravity causing the indefinite recession of a_n from $a = 0$ or as a force of attractive gravity causing the indefinite approach of a_n to $b = 1$ on the Euclidean line ab.

Because no matter how great is n of a_n, n is as far off from ∞ as the least finite point located at $a = 0$, it follows that a_n, similar to its index n, is as far off from $b = 1$ as the least finite point a_0 located at $a = 0$. This means not only is the indefinitely approaching a_n incapable of reaching its final point $b = 1$, but also that it is incapable of moving beyond itself—beyond the initial point $a = 0$. Motion along the Euclidean line ab is therefore impossible because the varying a_n can neither end nor even begin its acceleration.

The claim of empirical science that the objective of the observable universe or human society, which we designate by the finite variable a_n, is to indefinitely accelerate on the straight line ab is therefore a logical and physical impossibility—an appearance, or a scientific fantasy, dissimulating the absolute immobilization of the *finite variable* a_n—of the indefinitely accelerating human society—at the initial point a =0, and its incapacity to start any acceleration or progression whatsoever. It follows that the *real* objective of the varying human society is not indefinite acceleration aiming at accelerating up to the finite and sensible a_n, but rather the break of this sensible barrier and the maximum acceleration up to the real limiting point b =1.

Once human society reaches the real limiting point b =1, its rectilinear acceleration on the Euclidean plane at the indefinitely increasing speed a_n is transformed into real acceleration, which is circular acceleration on the infinite sphere of the physical universe. A body on the infinite sphere that partially closes the infinite Euclidean plane, has the power to move everywhere or anywhere on the unlimited Euclidean plane at one instant and at a maximum speed equal to the real speed of light c =1, which we defined as the product or ratio of infinite and zero speeds. We have called this instantaneous motion at a distance *infinite universal motion free of linear time.*

If the fundamental objective of acceleration is not indefinite acceleration, as contemporary empirical scientists (Ray Kurzweil and, Freeman Dyson, for example) claim, but rather maximum acceleration, then we can equally assert that the real objective of computation is not the indefinite increase of our brain's finite computational power; it is the attainment of our brain's maximum or infinite computational power, which is the end of computation. We have argued that the finite dimensioned measure of the brain's infinite computational power is roughly 10^{50}ops/s. As soon as the brain attains this real physical limit 10^{50}ops/s, which we take as infinite unity, and define as the product or ratio of infinite and zero magnitudes, the brain acquires infinite computational power that enables it to break the sensible barrier and hypercompute a_n, that is to say, compute the real infinite whole 1, by traveling and computing in one instant the infinitely many parts that separate a_n from b = 1.

This instantaneous action at an infinite distance is geometrically possible if we assume that the infinite distance separating a_n from its real limit b=1 is the zero distance, that is, the cosmic singularity, unifying a_n with b =1. Passing through this unifying cosmic singularity of zero extension and zero dimension allows us to travel and compute at once the infinite number of parts lying between a_n and b =1. This unifying cosmic singularity located at the convex part of the physical universe's limiting boundary b is, as we have argued, both outside our brain at the cosmological scale of $\approx 10^{26}$ meters and inside our brain at the quantum scale of $\approx 10^{-34}$ meters.

Now, insofar as the real number $1 = \infty/0$ computing the real limiting point b is a proportion between infinite and zero magnitudes, it is the end of the computable number and the emergence of the Platonic ideal number, which is both Nature's principle of the unity of opposites and a number computing the incomputable real limiting point *b* containing the universe's infinite totality of parts.

THE PARADOX OF THE BRAIN-EXTENDING TECHNOLOGIES

Another claim of technologically oriented scientists is that the acceleration and extension of our finite brain's limited powers are necessarily realized by external agents—machines. If *a* stands for brain and *b* stands for machine, we have the following logical inequality in which the techno-scientific vision of the world is grounded

$$a < b,$$

where the sign of inequality < designates inclusion (subsumption, subordination).

As it is indicated by the above relation of inclusion, the concept "brain" is included within the concept of machine such that every brain is a machine.

This logical inequality or conflicting divide is the result of the empirical assumption that our brain, like all other things in the Euclidean observable universe, is simple and uniquely finite. As a uniquely limited and incomplete being, the brain needs external causes, for example, a transcendent Mind, to create it and a transcendent Machine to sustain and improve it—to indefinitely repair and enhance its finite body. With the passage of time, this dependence of the brain on the machine progresses such that the machine's performance increases and, the brain's performance decreases up to the point where artifacts invade everything: our finite brain, society, and the observable universe. The indefinitely enhanced brain atrophies and eventually disappears, leaving its place to a new, superior intelligent species—the robot. Noted proponents of this idea are the transhumanist scientists Hans Moravec and Ray Kurzweil.

Now, the paradox with brain-enhancing technologies is that they precipitate the brain to extinction instead of improvement. The apparent extension of the brain through artifacts is in reality a reduction of the brain, which dissimulates the impossibility of extending the brain and in general human life by external agents such as transcendent machines. The reason is that the presence of external factors on which human life and the brain depend is based on external causality and temporal order, which are hierarchical principles of causal heteronomy and subordination characterizing dependent mortal beings conditioned by time. Given the fact that any improvement of the brain by external agents reduces the very proper power of the brain, as it makes it more dependent on these external agents, it follows that the real improvement of the brain must originate within the brain itself. Because real improvement is self-improvement, verifying the principle of immanent causality, or self-causality (causal autonomy, or free causality), we conclude that the founding principle of real brain improvement is not external causality but instead immanent causality, or self-causality.

THE PRINCIPLE OF COMPUTING NUMBER VERSUS THE PRINCIPLE OF INFINITE LIVING NATURE

The principle of self-causality (causa sui), which means "cause of itself," stipulates in its complete universality that everything has its cause within itself. It governs the absolutely necessary and free Being—namely, Physis—regarded as the embodiment of the Divine Being. An alternative expression of self-causality is the synthetic principle of self-order (reflexive order), or spontaneous order, which stipulates that everything is self-contained—that is, both a containing whole and a contained part—and is, therefore, greater than itself:

$$(a > a).$$

The principle of self-containment or self-order applied to material existence becomes the principle of self-existence, whereas applied to motion and mind (soul, consciousness), which we regard as immanent properties of material existence, it becomes, respectively, the principle of self-motion and of self-consciousness. Now, if we call the physical being, say the physical universe, having the power of self-order (self-existence, self-motion, self-consciousness), the living being— Plato, in his *Timaeus*, called the physical universe the *eternal animal*—then the above self-order is a principle of continuous life, in which continuous life is the power to be greater than itself, a containing whole and a contained part.

However, the analytic principle of computing number stipulating that nothing is greater than itself

$$(a > a)',$$

contradicts self-causality as self-order, and therefore life governed by the synthetic principle of self-order. Because the analytic principle of computing number destroys life, the dream of technological enthusiasts (Rodney Brooks, for example) to build computational models of living beings, whether these living beings are inorganic self-moving atoms or organic self-moving cells, is a pure contradiction—an impossibility. Finite analytic logic is not the Divine Logos of Physis, which is synthetic and indeterminate. In fact, the physical universe or nature is known exactly by analogy (proportion) and only approximately by computing number. Nature is the σύνθετon, that is, the composition of opposites, which Plato called *proportion*, we in modern terms call *ratio* $a/b = 1$ or equivalence $a = b$, and Aristotle called *syllable ab*.

In the synthesis, the distinct opposites are both affirmed and transcended (denied) in order to form an independent, intelligent and sublime whole. It is a fundamental contradiction in Aristotle's philosophical work that although he clearly recognized, and even emphasized, the synthetic nature of the physical being, the principles he postulated in his *Metaphysics,* which he regarded as the principles of being, are analytic and not synthetic.

By reducing the complex physical being into the sensible individual, Aristotle took what he observed with his individual senses—that is, the individual—*as if* it were the real physical being. He then took the analytic principles of the observable individual *as if* they were the real and permanent principles of the physical being, whereas in reality they are accidental rules determining our partial and individual perception of the physical being. We call *anthropocentrism* the above double illusion, perceptual and epistemological, caused by the imperceptible influence of our human individual perception on the physical whole from which Aristotle and all of us suffer.

RE-LINKING WITH OUR INNER INFINITE POWER SOURCE

Out there we have a complex physical being or whole, which we perceive *as if* it were a simple individual, and project the simple laws of the individual into the complex physical whole. It follows that what we observe is an approximation of the complex physical whole—of the real infinite whole 1. The issue is therefore to *re-link* ourselves with this highest and deepest reality—the real 1, which insofar as it is beyond our limited brain, is the real physical whole embracing everything, or, to put it another way, the physical universe's cosmic singularity unifying all oppositions and things.

Insofar as the real 1 is within our limited brain, it is the ultimate constituent of matter, our inner cosmic singularity or point-fire radiating infinite light energy and information (memory, consciousness, knowledge) at zero and infinite frequencies, at infinite and zero wavelengths. Re-linking with our inner infinite power source of energy and information will make us infinite brains in fusion with the infinite physical universe. This means we will have complete and exact knowledge of the physical universe, free of representation and approximation precisely because we replicate the infinite universe—we *are* the infinite universe. It follows that transcendent robots and other chimeric simulations, representations, and approximations of the living physical universe are useless.

As the infinite universe relative to our deepest reality, we have infinite universal sensation (hypersensation) that gives us omniscience, infinite universal motion that gives us omnipresence, infinite computational and memory powers that give us omnipotence. The rift between human and Divine Physis, between finite individual sensation and infinite universal thought, is finally healed.

FOR AN ACCOMPLISHED HUMANISM

The advancement of Plato over Aristotle is that Plato understood that due to the synthetic nature of Physis, or matter, the immanentization of the *eschaton*—the real 1—is of necessity in conformity with the synthetic principle of Logos or Physis: the unity of opposites. In this sense, we can rightly aspire to emancipate our human brain from its shadowy progression in the sensible world now, in the present, and not after death (transcendent theism) or after the human species (transcendent posthumanism). Because our human brain is both mortal and eternal, finite and infinite, inside the indefinitely progressing sensible world a_n and beyond it, at the real limiting point *b* of the physical universe—we have the power to transcend our indefinitely progressing sensible finite brain a_n and become *here and now* what we are since eternity "if the doors of perception were cleansed" (William Blake): that is to say, the real 1 designating the real infinite brain endowed with infinite life, motion, and consciousness.

Then and only then will we have *accomplished* our humanism, when we humans have actualized our distant infinity residing at the real limiting point *b*—the cosmic singularity—of the physical universe existing both inside and outside our human brain. If the actualization of our distant infinity governed by Logos—that is, by the synthetic unity of the opposites—is our final end that gives meaning to our acceleration, permanence to our chaotically changing lives, and constant truth to our indefinitely falsifiable theories, the feeling we experience from realizing our final end we call bliss.

We have reached here the real limiting point *b*, which tell us *where* our human society or our individual brain a_n accelerates. In fact, we accelerate to the limiting boundary *b* of the physical universe existing inside our brain at the scale of $\approx 10^{-34}$ m, and outside our human brain at the scale of $\approx 10^{26}$ m as the first origin, final end and unifying principle of the physical universe.

The limiting point *b* also tells us *why* we accelerate to the limiting boundary *b* of the physical universe. The answer is to complete ourselves, to become a real infinite whole 1 having the power to perform infinitely many operations in finite time, to travel the infinitely distant in one instant, and to sense the infinitely many as one. This instantaneous action at a distance free of time delay emancipates us from the pathologies of linear (irreversible) time and its correlative evils such as corruption, violence, injustice, subordination, comparison, contradiction and isolation. It also emancipates us from the malediction of the sensible universe's Euclidean sequence of cycles of generation and destruction, of explosion and implosion, that leads eventually to the exhaustion of its energy in conformity with the second law of thermodynamics, the law of linear time.

Finally, the limiting point *b* tells us *how* to accelerate to the limiting boundary *b* in a costless manner. We accomplish this by accessing our human brain's ultimate constituents—its inner cosmic singularities radiating energy at zero and infinite frequencies. Instead of building increasingly greater particle accelerators to attain increasingly higher levels of energy and temperature at an increasingly greater costs in terms of time, material resources, capital, and human effort, it is preferable to access our brain's inner sacred fire whose infinite source is, as we have asserted, its inner cosmic singularities.

In this way, by connecting the physics of high energy with the subjective experience of the infinite divine, the disenchanted observable world of our individual senses is re-enchanted.[1] The finite analytic paradigm of empirical science has disenchanted the world because it interprets the opposition and distance between any two spatial points *a* and *b* as a contradiction and temporal order. This in turn destroys the unity and coexistence of opposites and hence continuous motion between them.

We have demonstrated in this work, however, that, if we have the power to move from *a* to *b* and hence to be more than what we are at *a*, which is what life really is, it is precisely because we do not function according to the analytic principles of the finite analytic paradigm. In fact, we function according to synthetic principles—particularly according to the synthetic principle of the unity of opposites called the Logos, which governs our highest and deepest reality located at the limiting boundary *b* of the physical universe inside and outside our human brain and designated by the real 1. Thus, the paradigm that re-enchants the world is the infinite synthetic paradigm of rational science; it supresses time delay between *a* and *b*, here and there, the sensible and the real in order to restore their lost unity, coexistence and continuous communication.

This is not a magical world of mysterious influences at a distance but the enchanting and unifying Logos interconnecting any two parts *a* and *b* of the world regardless of their difference and distance in order to form a sublime comprehensive whole, which we call infinite nature or the real physical universe.

NOTES

1. The German philosopher Max Weber (1864-1920) used the term *disenchantment of the world* to describe the analytic computational paradigm of the modern world, which, in association with empirical science, eliminates the divine unity—the Logos—among the interdependent parts of the whole. The aim is to decompose the whole into a collection of isolated computable parts, the individuals, that verify analytic principles of organization. We have argued that the exclusion of this divine unity from multiplicity destroys continuous motion among the diverse parts and hence life itself. The world is disenchanted because it is inert, lifeless and Alogos, that is, without Logos.

The world is re-enchanted when the excluded divine unity—the Logos—is restored in order to transform the collection of isolated computable parts into a whole of interconnected parts, the incomputable living wholes.

Glossary

analytic principles. The tenets of the manner in which our finite individual senses perceive the physical universe. We call *finite analytic understanding* our cognitive faculty of analytic principles of organization. These are not principles of the real physical universe, but rather of our perception of it. In fact, our finite individual senses perceive the physical universe *as if* it were an individual thing or a sequence of isolated individual things without unity among them and which we call the sensible or observable universe. The analytic principles are: i) the principle of self-identity (reflexive identity), which states that everything a is equal to itself a: $a = a$; ii) the principle of contradiction, which states that nothing is both a and a' (not-a) called b: $(aa')'$, or $(ab)'$, or $(a = b)'$; iii) the principle of the excluded third, which states that everything is simple and determinate and hence is either a or b : $a + b$; iv) the principle of linear (irreversible) temporal order, which states that given any two different things $a \neq b$, either $a < b$ or $b < a$. Because on the Euclidean line opposite points a and $b = a'$ do not coincide, analytic principles forbidding the unity (or coincidence) of opposites are Euclidean principles.

contradiction. The impossible unity (or coincidence) of opposites. According to the analytic principle of contradiction, nothing is both a and not-a called b—for example, both limited and unlimited (not-limited): $(aa')'$, or $(ab)'$, or $(a = b)'$. It follows that contradictory opposites exist successively, that is, either $a < b$ or $b < a$. There is no common and intermediate element between contradictories (*see* also **analytic principles**).

contrariety. The necessary unity (or coincidence) of opposites (*coincidentia oppositorum*). According to the synthetic principle of the unity of opposites, everything is both a and a' (not- a) called b—for example, both limited and unlimited: aa', or ab, or $a = b$. Here the opposites are united (conjoined) in their mutual position or coexistence. We also have contrariety, when according to the synthetic principle of the included third, everything is neither (not-either) a nor $b = a'$—for example, neither limited nor unlimited: $(a + a')' = (a + b)' = a'b'$. Here the opposites are united (conjoined) in their mutual negation, or co-absence (*see* also **synthetic principles**).

de-coherence. The process by which our finite individual perception disrupts the unity and coherence of the whole by getting information about the whole *as if* it were a simple individual having at one time a unique determination and at different times consecutive determinations.

Euclidean geometry. The geometry of the sensible finite part—for example, of the sensible or observable universe, which we represent geometrically by the indefinitely varying (extending/dividing) Euclidean line or plane and designate by the finite variable a_n. The Euclidean observable universe is the inner, time-conditioned part of the real physical universe made of discrete simple things—the individuals—deprived of an intimate bond and verifying analytic principles or organization (*see* also **analytic principles**). Motion between any two points a and b of the Euclidean plane separated by an indefinitely varying distance is impossible, incomplete or apparent. The Euclidean plane is a space of absolute rest—a zero force space.

holographic principle (in extended form). The tenet that the maximum content (information or mass) contained in a specific spherical volume is proportional to its boundary area and not to its volume. The boundary (surface) area of the volume is proportional to the square of its radius.

individual. Anything that cannot admit opposite parts simultaneously. According to the analytic principle of contradiction no individual is both a and $b = a'$. It follows that if the individual is both a and $b = a'$, then the individual is self-contradictory, or paradoxical. The geometric form of the individual is the Euclidean point that has no parts and is thus a simple point without division, extension, and motion (*see also* **contradiction**).

infinite. The maximum. A quantity a is infinite if there is no other quantity a' greater than itself: $(a < a')'$. The infinite as maximum is that which cannot be further increased no matter how much we increase it. This refutes the analytic principles of contradiction, inequality and temporal order. In this sense, the infinite maximum is both constant and varying, closed and open, finite and infinite, whole and part, and thus reconciles all opposites. The infinite maximum, which we identify with the infinite whole, absolute infinite, or actual infinite, verifies synthetic principles of organization, such as that of the unity of opposites (*see also* **synthetic principles**).

The geometric representation of the infinite whole is the infinite sphere centered upon us and having an infinite radius. Seen from inside and along the Euclidean line of sight the infinite radius has infinite magnitude; seen from outside and at a right angle to the Euclidean line of sight, the infinite radius has zero magnitude. The complex product of infinite and zero magnitudes gives the radius $1 = \infty \times 0$, which is the indeterminate unit radius of the real infinite sphere taken as the body of the physical universe and numbered by the real infinite whole 1.

A quantity a is finite or potentially infinite if there is always another quantity a' greater than itself: $a < a'$. The finite or the potentially infinite has always something outside and beyond itself and obeys the analytic principle of inequality and temporal order. In this sense, it is the inner incomplete and time-conditioned part of the complete timeless infinite whole. The arithmetic representation of the potential infinite is the unlimited series of partial sums a_n without the limiting point and total sum $b = 1$ to complete and actualize the unlimited series a_n. On the other hand, the geometric 2-D representation of the potential infinite is the Euclidean plane a_n without the infinite sphere of zero radius to close and fixe it into an infinite whole. In this sense, the potential infinite verifies the analytic principle of contradiction, which stipulates that no unlimited series is limited and, that nothing is both unlimited and limited (*see also* **contradiction**). It follows that the unlimited series a_n and its contradictory limiting point b exist successively such that a_n is less and before b—$a_n < b$—and that the whole b is greater than any of its proper parts a_n: $b > a_n$.

The potential infinite or indefinite a_n is an apparent or improper infinite. In reality, it is a finite variable a_n varying without limit in an indefinitely varying Euclidean plane. Because of its indefinite variation in time, which generates the tension, frustration, and absurdity of infinite regression, the improper infinite is also an *evil infinite*. This evil infinite is transformed into a *good infinite* once the unlimited series of parts a_n is completed by its opposite, the limiting point and whole b, which captures and computes within an instant the sum total b of the unlimited series of parts.

We call actual infinite the unlimited series of parts a_n with a limiting point b that completes, fixes, and actualizes the infinite series. The actual infinite therefore refutes the analytic principle of contradiction (see above). In fact, it verifies the synthetic finite-infinite equivalence principle, which stipulates that everything unlimited is limited, and that the unlimited series a_n is equal to the limit b: $a_n = b$. It also verifies the synthetic whole-part equivalence principle, which stipulates that the whole b is equal to any of its proper parts a_n: $b = a_n$.

The actual infinite is the real or proper infinite—namely, the infinite as maximum, as the sum total of the unlimited series of parts—which we represented geometrically by the infinite sphere enveloping the Euclidean plane a_n. It symbolizes the complete whole: lacking nothing and therefore having nothing outside and beyond itself.

We can identify the sensible potential infinite a_n with our indefinitely accelerating observable world, which is the world of the time-conditioned partial sum of parts organized according to analytic principles. On the other hand, we identify the intelligible actual infinite, that is, the real infinite whole 1, with God or with the real physical universe taken as the physical body of God. The real physical universe 1 is the totality (sum total) of the unlimited number of parts a_n organized according to synthetic principles.

Among philosophers and mathematicians who considered the finite or potential infinite to be the only infinite in conformity with our finite brain's analytic principles of organization and rejected the actual infinite are: Aristotle, Kepler, Descartes, d'Alembert, Kant, Gauss, L. Kronecker. . . . Those who defended the existence of the actual infinite or infinite whole or absolute infinite include: Anaximander, Melissus, Pythagoras, Plato, Nicholas of Cusa, Bruno, Pascal, Spinoza, Newton, Leibniz, Blake, Bolzano, Hegel, Dedekind, Cantor. . .

limit. i) The extremity of a thing: that is, the first point beyond which it is not possible to find anything and the first point within which all points reside; ii) the primary being; iii) the end of anything: that to which, not from which, a movement or action proceeds; iv) the origin or unifying principle (ἀρχή); v) the form of that which has magnitude: the body; vi) the what *(quiddité)* of anything, which is the limit of knowledge but also the limit of the known thing; vii) the place in which something is something else (the complex limit).

Non-Euclidean geometry. The geometry of the physical infinite whole in which the Euclidean analytic principles of contradiction, inequality and linear (irreversible) temporal order are refuted.

synthetic principles. The founding principles of the real physical universe. We call *infinite universal reason* the cognitive faculty that apprehends the physical universe governed by synthetic principles of organization. Synthetic principles are: i) The principle of the unity (or coincidence) of opposites, which we call synthetic reason, Logos, Physis; its formal expression is the synthetic principle of equivalence, which states that everything a is equal to something else $b = a'$: $a = a'$, or $a = b$, that everything is complex and hence both a and $b = a'$: aa', or ab, and that opposites are therefore coexistent; ii) the principle of the included third, which states that everything is indeterminate and impartial and hence neither (not-either) a nor b: $(a + b)' = a'b'$, and that opposites are therefore co-absent; iii) the principle of non-linear (reversible) temporal order, which states that given any two equivalent things $a = b$, then both $(a < b)$ and $(b < a)$; iv) the principle of self-causality or self-order, which states that everything is a cause of itself: $a < a$, that everything is greater than itself: $(a > a)$.

Because on the geodesic of the infinite sphere opposite points a and $b = a'$ coincide despite their maximum or infinite opposition, synthetic principles affirming the unity of opposites are *spherical principles*. There is a common intermediate and unifying element between contraries.

universe. Anything that admits opposite parts at the same time, which we call contraries (*see also* **contraries**). According to the principle of self-order or self-containment the universe is both the enveloping whole and the enveloped part: $a > a$. According to the synthetic principle of the unity (or coincidence) of opposites, the universe is a complex thing that is both a and $b = a'$, limited and unlimited, one and infinitely many: aa' or ab. Indeed, from the Latin words *unus* or one and *versus* (inverse), the word *universe* means both one and the inverse of one, which is the infinitely many. According to the synthetic principle of the included third, the universe is an indeterminate thing that is neither (not-either) a nor $b = a'$, neither limited, nor unlimited, neither one nor infinitely many: $(a + b)' = a'b'$. Ultimately, the universe is free of the determinate part by comprising the totality of determinate parts (*see* also **synthetic principles**).

We may represent the universe geometrically by Plato's point-line, which is both an undivided point and an infinitely divided line—that is, one and infinitely many. The universe as point-line emerges from the balance of equal and opposite forces—of the shrinking force of attraction, which generates the point of infinite curvature, and the stretching force of repulsion, which generates the line of zero curvature. The balanced ratio of infinite and zero curvatures determines the curvature $1 = \infty/0$, which is the complex indeterminate unit curvature of the universe represented geometrically by the infinite sphere or point-line.

Select Bibliography

Aristotle. *Physics*. Translated by Henri Carteron. Paris: Les Belles Lettres, 1983.

Aristotle. *Metaphysics*. Translated by J. Tricot. Paris: Librairie Philosophique J.Vrin, 1984.

Aristotle. *Ethica Nicomachea* . Translated by W. Ross. London: Oxford University Press, 1975.

Aristotle. *On the Heavens.* Translated by J. Tricot. Paris: Librairie Philosophique J. Vrin, 1996.

Aristotle. *Organon.* Translated by J. Tricot. Paris: Librairie Philosophique J. Vrin, 1984.

Aristotle. *On the Soul.* Translated by J. Tricot. Paris: Librairie Philosophique J. Vrin, 1992.

Brooks, Rodney. "Beyond Computation." In www.edge.org

Cantor, George. "*Fondements d'une théorie générale des ensembles.*" In *Mathematische Annalen* XXI 1883, and in *Cahier pour l'Analyse: "La Formalisation.*" Paris: Editions du Seuil, 1969.

't Hooft, Gerard. "*The Holographic Principle.*" In arXiv:hep-th/0003004v2, 1 May, 2000.

Dyson, Freeman. *Infinite In All Directions*. New York: Harper & Row Publishers, 1988.

Kant, Immanuel. *Critique of Pure Reason*. London: Everyman's Library, 1991.

Kirk, G.S., Raven J. E., and Schofield, M. *The Presocratic Philosophers*. Cambridge: Cambridge University Press, 1988.

Koyré, Alexandre. *Du Monde Clos A L'Univers Infini.* Paris: Press Universitaires de France, 1962.

Kurzweil, Ray. *The Singularity Is Near: When Humans Transcend Biology*. New York: Penguin Books, 2005.

Lloyd, Seth. "*Ultimate Physical Limits to Computation.*" In arXiv:quant-ph/9908043v3 14 February, 2000.

Moravec, Hans. "*When Will Computer Hardware Match the Human Brain*." In Journal of Evolution and Technology 1, 1998.

Nicholas of Cusa. *On Learned Ignorance.* Translated by J. Hopkins. Minneapolis: The Arthur J. Banning Press, 1985.

Pascal, Blaise. "*Les Pensées*." In *Oevres Complètes*. Edited by Jaques Chevalier. Paris: Bibliothèque de la Pléiade, 1954.

Plato. *Philebus.* Translated by E. Chambry. Paris: Flammarion, 1969.

Plato. *Timaeus*. Translated by E. Chambry. Paris: Flammarion, 1969.

Russell, Bertrand. *The Principles of Mathematics*. London: George Allen & Unwin, 1979.

Susskind, Leonard. "*The World as Hologram*." Lecture, 2000.

Weber, Max. "*Science As A Vocation*." In *From Max Weber: Essays In Sociology*, edited by H.H.Gerth, and C.Wright Mills. London: Routledge § Kegan Paul Ltd, 1947.

Index